THIRTY WAYS OF LOOKING AT HILLARY

THIRTY WAYS OF LOOKING AT
HILLARY

REFLECTIONS BY WOMEN WRITERS

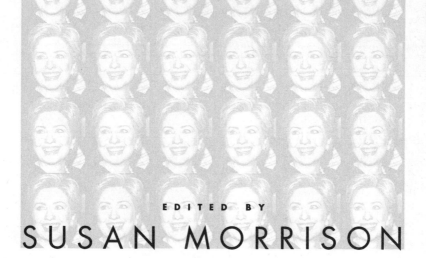

EDITED BY

SUSAN MORRISON

HARPER

An Imprint of HarperCollins*Publishers*
www.harpercollins.com

FIRST EDITION

Designed by William Ruoto

Library of Congress Cataloging-in-Publication Data is available upon request.

ISBN: 978-0-06-145593-3

08 09 10 11 12   ID/RRD   10 9 8 7 6 5 4 3 2 1

FOR HELEN AND NANCY

CONTENTS

In an episode of *The Simpsons* from a few years ago, Bart gets a glimpse thirty years into the future and sees that his sister Lisa has been elected president. The show doesn't make a lot of fuss over the notion of a woman commander in chief. (In fact, in an aside we learn that Lisa is "the first *straight* woman president.") But it's impossible not to notice that President Simpson looks a lot like Hillary Clinton: her spiky hairdo has been smooshed down into a power-helmet; she wears monochrome pantsuits and a string of pearls. She is intelligent and earnest, and she has a problem with authenticity. Bart, who has grown into a middle-aged slacker with a ponytail, ends up playing Bill to Lisa's Hillary; he saves the day by smooth-talking a mob of angry world leaders who have converged on the White House. At one point he says, "I coulda been president, but I'm too real."

Authenticity is shaping up to be the buzzword of the 2008 presidential campaign. Americans always like to believe that they really *know* the candidate they are voting for. And although she is probably the most famous woman in the world right now, Hillary Clinton has a lot of people stumped. It could be that we think we should know her well already, having watched her and her husband for eight years in the White House (and having learned more about their marriage than we had a right to, thanks to Kenneth Starr). Or it could be that, because she is a woman, we have different expectations of her and how cozy we ought to feel with her.

On a shelf in my kitchen is a campaign button that I picked up during the 1992 presidential race. Over a photo of Hillary (bangs and headband phase—which was basically my look then, too) are the words "Elect Hillary's Husband." Back then, the slogan produced a kind of giddy frisson: not only was the candidate just like someone I could have gone to college with—a baby boomer—but his wife was, too. And she had a job! I had only known first ladies as creaky battleaxes who sat under hairdryers and wore brooches. The thrill associated with that button feels far away now, and it's hard to know exactly why. There's no doubt that the rinky-dink scandals of the Clinton administration and the dismal parade of special prosecutors took the gleam off the fresh start that the Clintons brought with them to Washington. But that doesn't quite explain how now, fifteen years later, there is not more simple exuberance at the idea that we may be about to elect our first woman president.

Walter Shapiro, writing in *Salon*, suggested that it won't be long before some liberal arts college creates a department called Hillary Studies. He was referring to the way biographies of the former First Lady keep piling up, but he might just as well have been talking about the amount of time people spend scrutinizing her. We don't seem to discuss Hillary Clinton the way we discuss other politicians: we raise our voices, we argue, we wave wine glasses around at dinner parties.

"I'm a Rorschach test," Clinton herself once said, referring to the way people tend to project their own hopes and anxieties onto her. To some, Hillary is a sellout who changed her name and her hairstyle when it suited her husband's career; to others she's a hardworking idealist with the political savvy to work effectively within the system. Where one person sees a carpetbagger, another sees a deft politician; where one sees a humiliated and long-suffering wife, another sees a dignified First Lady. Is she tainted by the scandals of her husband's presidency or has she gained experience and authority from weathering his missteps? Cold or competent, overachiever or pioneer, too radical or too moderate, Clinton continues to overturn the assumptions we make about her.

No other politician inspires such a wide range of passionate responses, and this is particularly true among women. As I talked with women about their reactions to Hillary, some themes came up again and again. Many women were divided within themselves as to how they feel about her, and I noticed a familiar circle of guilt: these women believe they should support Hillary as a matter of solidarity. But because they expect her to be different from (that is, better than) the average male politician, she invariably disappoints them; then they feel guilty about their ambivalence. Some feel competitive with her. Having wearily resigned themselves to the idea that "having it all" is too much to hope for, they view Hillary as a rebuke: how did she manage to pull it off—or, at least, to appear to pull it off? Other women say they *want* to like her but are disturbed by the anti-feminist message inherent in the idea of the first woman president getting to the White House on her husband's coattails. Then there are women, like the late playwright Wendy Wasserstein, who are queasy over the way Clinton's popularity spiked only after she was perceived as a victim. When it became clear that Hillary was going to stand by her man after the Lewinsky fracas, Wasserstein wrote a disheartened op-ed piece in the *New York Times*. "The name Hillary Rodham Clinton no longer stands for self-determination, but for the loyal, betrayed wife," she wrote. "Pity and admiration have become synonymous."

There's plenty of Hillary Studies literature out there that parses the candidate's stands on policy issues, her Senate votes, and her track record as First Lady. This book isn't aiming at that kind of op-ed page territory. Rather, it's an attempt to look at the ways in which women think about Hillary (and why they think so *much* about Hillary), how they make their judgments about her, which buttons she pushes in them and why. It aims to pull together some of the many refracted visions of Hillary.

The defining events—some might call them bloopers—of Clinton's political progress have become well-worn touchstones, and, not surprisingly, some of them turn up repeatedly in these pages. There was her widely misunderstood remark about Tammy Wynette, her mutating hairstyles, the matter of her maiden name, the botched health-care initiative, "cleavagegate," and, most famously, the cookies-and-tea brouhaha. (It's interesting that when Laura Bush made her own baking-related remark, in the fall of 2007, not a single reporter followed it up: assailed for phoning the UN secretary-general personally to ask him to denounce the junta in Myanmar, she said, "This is sort of one of those myths: that I was baking cookies and then they fell off the cookie sheet and I called Ban Ki-moon." Maybe we've come further than we thought since 1992.)

There is a healthy bit of cross talk in the pages that follow: some of the writers argue among themselves about Hillary. It's all in the nature of a discussion. Clinton announced her candidacy in a videotaped statement showing her seated on a sofa in her living room. As she said in the video, "Let the conversation begin."

THIRTY WAYS OF LOOKING AT HILLARY

YELLOW PANTSUIT

WHAT IF YOU COULD NEVER SAY WHAT YOU REALLY FELT?

BY AMY WILENTZ

My generation definitely has a Clinton problem. The problem consists in what psychiatrists call transference: the men and women of the baby boom take what we imagine ourselves to be, all the myths and ideas and emotions of our cultural and political selves, and then we project it onto the Clintons. What a burden for them, I often think. But then, they *asked* for it.

We can't help but identify with these two bizarre creatures—the grand Democratic figures of our adulthood, the Mom and Pop of the Me generation—even though the creatures were formed in Arkansas. Frankly, there are no other Arkansans with whom I've ever identified. I can't say I've ever met a person from Arkansas, except for Bill Clinton. Even though he was president, he seemed to typify the Arkansan breed in those few seconds I shared with him: he used the words "might" and "could," one right after the other, something I'd never heard before outside the Li'l Abner comic strip.

Do you think things will turn out well for Haiti, I asked him.

"Well," said the president thoughtfully. "They might could."

Our Clinton problem now, of course, is Hillary (for the moment, Bill's only a useful accessory, like Barbie's Ken, except Bill has that extra part that sometimes gives trouble). Just as Bill raised issues for men of our generation about what kind of man one should be, what kind of leader, husband, father, etc., so Hillary raises similar issues for women.

Personally, when men are running for office, or have been elected, I don't wonder what it's like to be them; or not very often. I have to say that I did wonder what it could possibly be like to be George Bush when he was sitting in a schoolroom reading about a goat while the World Trade Center was falling down. I did wonder why Howard Dean made that funny noise, I wondered about Bill Clinton's cigar wielding, and I did wonder about Ronald Reagan's inner life. Male politicians have their moments of head-spinning eccentricity that make you curious. Usually, though, it's an abstract humanity that one shares with balding, compressed-lip men, in suits, ties, and wing tips.

But Hillary is different. From the day she appeared on the national scene, I've been identifying, or at least trying to identify, with her, her goals, and most of all, her character (the hardest thing to pin down). I have always wondered what it would be like to be her, even though the "her" I'm wondering about has certainly changed over the past fifteen years.

Sometimes, my instinctive desire to identify reminds me of childhood fantasies about inappropriate or silly exemplary figures. Like many girls of ten, I would lie on my bed and wonder *so hard* about what it would be like to be Nancy Drew with all those adventures, climbing up a ruined mansion's chimney, driving around in the red roadster. I would wonder *so hard* about what it might feel like to be Jackie Kennedy, a beautiful and romantic widow. I would imagine *so intensely* being Clara Barton and wiping the feverish brow of a wounded soldier. There was something about each of these figures, I felt, that was just like me.

With Hillary, that feeling is more intense but less romantic. From the beginning, I looked at her and I couldn't help thinking that in many ways, I am like her and she is like me. I'm a Democrat and so is she. I'm a liberal and so—arguably—is she. We're both middle-class suburbanites. She's only a few years older than I. Okay, she's from Illinois, but is that so different from my home state, New Jersey? Illinois is working-class, a labor state, on America's hidden third coast (the Great Lakes), and as blue as they come. And I was amused and impressed by what Hillary said about baking during the first Clinton presidential campaign in 1992: "I've done the best I can to lead my life. . . .You know, I suppose I could have stayed home and baked cookies and had teas, but what I decided to do was fulfill my profession." That resonated.

I remember feeling delighted when she said that. I thought: there's our voice. The statement had a tone to it: arrogant, dismissive, irritated, ambitious, dedicated. She got ripped for it, of course: What's wrong with baking cookies, Hillary, huh? Are you too good for that? But she did think she *was* too good for that, and so did hundreds of thousands of us. We were made for things that were better than baking. It was before women decided it was better to leave the workforce and be *good mothers*. (In a post-feminist apologia, may I add that I have become a fairly accomplished bread baker—from scratch, without a machine—and I have also baked three children's birthday cakes per year for going on twelve years now. You *can* have it all.)

In fact, my apologia is very much in the style of the revised Hillary. Much to my dismay, I discovered, when I Googled "Hillary baking," that she has actually managed to get her name associated with a recipe for chocolate-chip cookies (one that uses oatmeal and a cup of vegetable shortening and that sounds awful). In 1992, after the infamous "baking" comment, Hillary—who rarely takes a step forward without taking two back, we now know—entered this cookie formula into a *Family Circle* magazine bake-off against other candidates' wives' recipes. She won. Of course. It is reasonable to assume that she did this in order to vitiate the original baking comment that stirred so much right-wing anger.

This to-and-froing is the kind of thing that has led women to wonder about Hillary, and that keeps them riveted. Her early, less self-conscious, less visibly political behavior was a confidence deposit that has—for me at least—not really depreciated, in spite of her more recent waffling and backpedaling. Back in the old days when she was honest, she said that thing about baking. She made fun of Tammy Wynette's "Stand By Your Man." I saw a picture of her with Chelsea in 1983, standing in front of the refrigerator in their kitchen. Chelsea was a toddler and Bill Clinton was the governor of Arkansas. Hillary was wearing semi-bohemian clothes—a long skirt and boots and a baggy shirt, and her hair was down in front of her face. She seemed irritated with Chelsea, who looked as if she was behaving "defiantly," as we say now. In another early picture, Hillary and Bill are looking down at the newborn Chelsea, and again, Hillary's hair is a wreck and she's wearing a soft headband and those big, tortoiseshell glasses, heavy and unfashionable, with which I feel so at home. The photographs reveal an earlier Hillary Clinton—Hillary Rodham, in fact—and a person who seems more like Howard Dean's publicity-shy doctor-wife Judy, and less like . . . Hillary Clinton today.

I like to believe that Hillary is still like the old Hillary Rodham, the pre-presidential Hillary. That behind the blond helmet lurks the bad brown hair and the shoved-on headband; behind the contacts or laser surgery she's looking out at the world through insectlike glasses. This brings me back to that original childhood fantasy: What must it be like to be her? There she is, retooling herself for America. She seems to have worked on her public character like a method actor until it's almost a part of her. But as I sit in front of my computer, watching a clip of her running up onto a dais in her yellow pantsuit during the summer of 2007, I can't help but wonder how much time and money was spent focus-grouping her wardrobe and whether she really wants to look like that. What is the real Hillary wearing in her own mind? Is the yellow pantsuit something that's begun to feel natural to her, or is it a costume?

Something about Hillary Clinton makes her seem uncomfortable in her new disguise. That something, which I recognize (and salute!) is an inability to choose a feminine role that feels comfortable to her, a trait she shares with many, but not all, professional women of her generation. Among prominent women, Hillary seems least at home in her outward presentation. She has none of Ségolène Royal's seductive, graceful charm, nor is she in the old-fashioned, sexless-grandma mold of Madeleine Albright. In part this is personal, but in part, it's because her generation lies smack in the middle of these two: Albright b. 1937; Clinton, b. 1947; Royal, b. 1953. Hillary is from the changeover generation, who were raised by lovers of pink and lace, and who wore overalls and jeans and miniskirts, went bra-less in their Wonder years, and then settled into a macédoine of fashions as they moved into middle age.

This latest edition is a Hillary who has done herself up for the general election. She's not aiming at me. Most of the time, she looks like a Republican. She gives off something between a country-club, golf-playing, hedge-funding vibe, with a whiff of bingo games, Sunday churchgoing, supermarket aisles, and coffee klatches. Her target is obviously some well-imagined political center. But remaking yourself in the national mold is no simple task. Not only does it take a lot of high-paid advisers (remember Naomi Wolf and Al Gore's "earth tones"?) but it demands an odd combination of self-abnegation and self-reinvention.

What would it be like, I wonder, to give myself that look while remaining—essentially—me? Is it possible? Style has meaning, as any woman will tell you, and any anthropologist. There's a difference in meaning between what Tom Hayden wears when he hangs out with gang members in downtown L.A. and what Donald Rumsfeld wears to board meetings, between what Paul Farmer wears to work at his health center in Haiti and what George Bush picks out of his closet for meetings down at Crawford. This is why Bush's flight suit and Hillary's cleavage on the Senate floor were so significant. The meaning of everyone's style is absorbed reflexively by his or her friends and acquaintances, and the significance of any public person's style is parsed by voters and viewers, even when they don't realize it.

A person usually chooses his or her own clothing and furnishings in an attempt, often vain, to express something about his or her character, and to be appealing in some way. I sympathize with anyone who has to make those choices based only on the taste of others, who has to subsume her own instincts in those of a committee (even a committee hired by herself), who has to appeal to the broadest possible audience, instead of simply trying to appeal to an ideal man, or an ideal professional circle, or—in the worst case—her own mother's taste. Every day, Hillary Clinton, besides having to be a real person in her own mind, has to placate and appease a broad spectrum of America. Every day, instead of living in her own eccentric personal familial world (as the rest of us do), she has to live in front of all America. Imagine waking up and knowing that on that day, as on every other, you have to pretend. You *want* to pretend. What if every morning, you woke up thinking about how you—a *Jeopardy*-watching, Chaucer-reading, French-speaking, TM-practicing American, possibly, or whatever Hillary is—had to conform to perceived national norms?

What if you had to operate in a universe where you were never allowed to say what you really felt? Could you go out and buy foie gras at a specialty market if you were running for president? Could you admit that you spoke French (John Kerry did, fatally)? Could you drink a Starbucks latte in public? Could you curse? Or smoke? Could you, as I often do, miss three consecutive appointments to get your hair cut? And really: what if you had to wear pantsuits or a turquoise jacket with a turquoise necklace and turquoise earrings? How would you explain this to your real friends? To say nothing of trying to explain, to people with whom you secretly agree, the reasons why you went against what they, and *you*, believe when you initially came out for the war in Iraq.

The reason that Hillary Clinton's outfits—her costumes—are important is that they seem of a piece with her ideological getup. It's easy to imagine, given what we know about what went before, that the two (her yellow pantsuit and her vote on Iraq) were insincere, meant as crowd-pleasers.

Worse is to imagine that both are heartfelt.

2

THE TYRANNY OF HIGH EXPECTATIONS

On doing whatever it takes

BY ELIZABETH KOLBERT

Hillary Clinton's career as a politician, rather than as the wife of one, began, more or less officially, on July 7, 1999, the day she announced that she was forming a campaign committee to run for the U.S. Senate. As the site for the announcement, Clinton chose Daniel Patrick Moynihan's farm in the upstate New York town of Pindars Corners. She conferred with Moynihan, who had never particularly liked her, or for that matter her husband, in a little one-room schoolhouse that he used as an office. Then she walked with him to a newly mown hay field where risers had been set up for the hundreds of reporters who had come to record the spectacle, some from as far away as Japan. Clinton pronounced herself "very humbled and more than a little surprised to be here." She raised the questions that, she supposed, were on everybody's mind—"Why the Senate? Why New York? And why me?"—but left them hanging, unanswered, in the pollen-filled air.

That same day, Clinton embarked on her famous "listening tour." Her first stop was Oneonta, where five citizens had been selected to share their concerns. A local doctor spoke about the importance of public education. A college student offered his reasons for wanting to become a teacher, and a businesswoman described the difficulty of finding good help among the local young people. "The one problem that I've noticed throughout all the years I've been in business in Oneonta is social skills and emotional intelligence," the woman said. "The problem of people answering the phones, unbelievable! I say to them, if you cannot relate in this little town, what will happen when you get out there and really see what the world is and don't have someone like me for a boss that takes you in the back and coddles you." It was hard to get a good view of Clinton, since most of the reporters from the risers were now crammed into one room, but every time she turned my way she was, it seemed, nodding emphatically, fairly radiating earnest concern. She was concerned about public education. She was concerned about teaching. She was concerned about access to dental care and better markets for dairy products and, of course, about the problem of answering the phone. Half an hour into the ninety-minute session, I counted three reporters fast asleep.

I found the listening tour difficult to take, and not just because it was sleep-inducing. The truth of the campaign—already obvious at that early date—was that New York was just a vehicle for Clinton's ambition. ("If she wins in New York, it will only be a matter of time before she announces for president," William Powers, then New York's GOP chairman, warned.) Yet here she was, traveling around the state, acting as if her carpetbagging was a reflection of her heartfelt concern for the views of Oneontans. The logic of the exercise was circular or, in its maddening topology, perhaps more like a Mobius strip. Clinton argued that she was justified in running from a state that she had never lived in because what mattered was "where you stand, not where you are from." But when asked where she stood, she kept insisting that she had come to New York to listen. "All I can say is, I care deeply about the issues that are important in this state that I've been learning about," is how she put it, absurdly, that first day.

At the same time, I was bothered by the fact that I was bothered. Sure, the listening tour was bogus. Sure, Clinton was disingenuous. So were Rudy Giuliani and George Pataki and Ed Koch and Mario Cuomo and David Dinkins and every other politician I had ever covered in New York, or any other state. The big lie is a classic campaign tactic, and Bill Clinton, as much as anyone else, was a master of the strategy. But with male politicians, a certain baseline duplicity is simply taken as a given.

Women come to any political race with all sorts of handicaps. Either they're not tough enough to be in charge, or they're too bitchy to be. Either they're too masculine or not masculine enough. If they're good looking, it's held against them; if they're not good looking, same deal. Being married and having kids is a problem, as is being unmarried and childless. Finally, there's the implicit double standard that goes with expecting more of them: we assume that female politicians will be purer of motive than their male counterparts, more driven by empathy and by issues than by personal ambition. (Congresswoman Carolyn McCarthy, of Long Island, whose husband's murder led her to campaign for stricter gun control laws, which, in turn, led her to run for office, is the very model of a model female candidate.) In short, we expect them to be above politics. To borrow (and reverse) a phrase from George Bush, this might be called the tyranny of high expectations. When you think about it, it is at least as debilitating as assuming that women are less capable.

Pretty much by definition, politics is about power. Power is rarely gained through civility, or niceness, or purity of heart. It has to be wrested from those who already have it and who are, as a rule, loath to give it up. If Clinton weren't a woman she could quite easily be admired for her remorselessness. Like the legendary Tammany Hall boss George Washington Plunkitt, she truly can boast, "I seen my opportunities and I took 'em." When, in 1992, Gennifer Flowers almost sank Bill's candidacy, Hillary rescued it; she then turned around and, as payback, demanded to be put in charge of the president's health-care task force. Her failure in this position was so colossal it would have prompted almost anyone to swear off politics forever. Hillary merely began plotting her come-back. When Monica Lewinsky, of all unlikely people, boosted Hillary's poll numbers, she decided to run for Senate. As it happened, she reached this decision just as her husband was being impeached! (So much for empathy.) At every one of these junctures, clear-sighted Clinton watchers saw through her poses—"Lady Macbeth in a black preppy headband" was Maureen Dowd's memorable formulation. But Clinton has always got, in a manner of speaking, what she came for. She has now been the junior senator from New York—a state that she never did live in—for nearly eight years, and although she has not taken any particularly bold stands or accomplished anything particularly significant, she has more than held her own. When she ran for reelection, in 2006, the Republicans could barely even muster a candidate to run against her, and she won with 67 percent of the vote.

How are women going to make it to the top? How are they—we—going to rise to the highest levels, not just in politics but in law, medicine, and hedge fund management? (According to the most recent figures, women head just twelve Fortune 500 companies, which comes to fewer than one in forty.) One theory has it that women will eventually triumph by virtue of their innate talents; as Charlotte Whitten, the first female mayor of Ottawa, famously put it: "Whatever women do they must do twice as well as men to be thought half as good. Luckily, this is not difficult." Another theory has it that as more women attain power—however they manage to do this—perceptions will change, so that traits like, say, cooperativeness will gradually come to be more highly valued than attributes like, say, aggression. By this account, the goal of women in positions of power should be to change the way power operates.

Hillary represents or, if you prefer, embodies a third alternative: exploit the prejudices against you. If people think women are good listeners, well, then, stage a listening tour.

Over and over again in her career, when Clinton has been attacked for being too ambitious, her response has been an elaborate show of femininity. When, for example, she came under assault for meddling with national health-care policy, she granted an exclusive interview to a *New York Times* food writer and posed for the accompanying photo in a shoulder-baring gown. (The *Times* indulgently ran her views on the White House menu on the front page.) Even as she was running for Senate, Clinton was penning—or at least putting her name on—*Invitation to the White House*, a lavish coffee-table book that included recipes for a cherry yule log and herbed tomato frittata. And when she arrived in the Senate, she impressed her (mostly male) colleagues with her winning deference.

"I had seen her a few times through a glass darkly, as the Scripture says," Senator Robert Byrd, who had tangled with Clinton over health care, told me. "But she is one of my favorites now. Because I like her approach. I like her sincerity. I think she has been the perfect student."

Running for president is the consummate expression of Hillary's ambition—as it is for her too-numerous-to-mention rivals. Still, as she opened her campaign, Clinton could hardly have been more self-deprecating. In her video announcement, which she posted on her website, she was sitting on a couch, as if she had just stopped by for a cup of coffee. "I'm not just starting a campaign," she said, smiling for the camera. "I'm

→ beginning a conversation—with you, with America. Because we all need to be part of the discussion if we're all going to be part of the solution."

→ "So let's talk," she went on. "Let's chat. Let's start a dia-logue about your ideas and mine. Because the conversation in Washington has been just a little one-sided lately, don't you think?"

That Clinton would engage in such a charade doesn't make one admire her. Yet one *has* to admire her. What is hardest to take about her and what accounts for her success are basically the same thing: she is willing to do whatever it takes. Women should wish for a more principled candidate. They should wish for one who's more honest. But they shouldn't expect to find her any time soon.

3

HILLARY ROTTEN

BY KATHA POLLITT

Ice Queen. Hag. Witch. Bitch. Dumb bitch. Feminazi. Slut.
Slut?

Type just about any misogynist insult into Google with "Hillary Clinton" and up will come hundreds of thousands of links. Promiscuity might not be the first sin that comes to mind when you think of her, but last time I checked " Hillary Clinton + slut" brought up 208,000 results. True, a lot of these are aimed at her husband or his various women friends, but prim and proper Hillary gets plenty of her own, as in "I will never vote for that corporate slut," and "HILLARY CLINTON IS A SLUT."

To you, to me, to a professional political analyst, Hillary Rodham Clinton may be a wonkish centrist Democrat, a believing Methodist, a people-pleaser, a trimmer. But to a lot of people, especially men who spend a lot of time online, she taps into some deep Jungian unconscious well of evil female archetypes: she's Snow White's evil stepmother, Jezebel, Lady Macbeth, Marie Antoinette, and Nurse Ratched all rolled into one. In other words, she's a powerful liberal woman. An _older_ powerful liberal woman. An older powerful liberal woman whose power is illegitimate because it is bound up somehow with sex—how else could a woman get power over men, its rightful possessors? In a country where it is still controversial for a married woman to keep her name—something Hillary Rodham was unable, in the end, to do without making her husband look henpecked in the eyes of Arkansas voters—women with power are automatically suspect. Even on the supposedly PC left, prominent right-wing women attract far more vitriol than comparable men, and of a more personal nature—remember Katherine Harris and her makeup? Or the rumors that she slept her way into political office? But a powerful woman who is perceived as liberal, perhaps even feminist, awakens at many different points on the political spectrum a kind of free-floating late-night hysteria that would be funny if it didn't have real-life consequences, like making a lot of smart people think that Hillary Clinton is unelectable.

Robotic. Angry. Dragon lady. Castrating. Lesbian. Cunt. Pig. Hillary Rotten, daughter of Satan.

It isn't just anonymous nuts who talk like this. Don Imus referred to Hillary as Satan constantly—"that buck-tooth witch, Satan"; Bill Clinton's "fat, ugly wife, Satan"—eleven times on just one show. CNN's Glenn Beck called her "the Antichrist." Michael Savage called her "Hitlerian." Chris Matthews called her "sort of a Madame Defarge of the left." Rush Limbaugh, who devoted many hours of radio raving to floating the charge that Vince Foster had not killed himself but had been murdered at the Clintons' behest, described her as "the woman with the testicle lockbox," whatever that means. "When she comes on television, I involuntarily cross my legs," said CNN's Tucker Carlson, in a jokey segment about the Hillary Nutcracker, which crushes walnuts between its steely thighs, yours for only $19.95.

The ball-busting theme looms large in the male Hillary-hating imagination—that's why she can be both a lesbian *and* a siren who has Bill by the short hairs. But it isn't just men who get a bit unhinged by Hillary. Maureen Dowd has compared her to mafia killer Tony Soprano, "so power-hungry that she can justify any thuggish means to get the prize." Peggy Noonan: "Cold and ambitious." Ann Coulter: "Pond scum." "White trash." Well, okay. For Ann Coulter, that's fairly restrained.

In the ordinary way of things, I probably wouldn't vote for Hillary Clinton in the primary. I'd vote for John Edwards on straight policy grounds, or for Barack Obama as the slightly lefter, fresher face, the candidate who can bring in the most new voters and give the United States' relations with the rest of the world a new start. Or I might just say the heck with the top three—who are much less different from one another than their self-branding would have you believe and who would probably do equally well (or not) if elected—and vote for Dennis Kucinich, who actually stands for things I believe in, like single-payer health care and not bombing Iran. But then I come across one of these sulfurous emanations from the national collective unconscious and I want to sit down and write Hillary's campaign a check *immediately*. I want to knock on doors for her every Saturday from now until primary day, on which I want to vote for her—*twice*. Sisterhood is powerful! We are all Hillary Clinton!

After all, the hysterical insults flung at Hillary Clinton are just a franker, crazier version of the everyday insults—*shrill, strident, angry, ranting, unattractive*—that are flung at any vaguely liberal, mildly feminist woman who shows a bit of spirit and independence, who puts herself out in the public realm, who doesn't fumble and look up coyly from underneath her hair and give her declarative sentences the cadence of a question. I've found a Hillary-with-a-penis photo in my in-box more than once, to say nothing of anonymous e-mails calling me a "fat, ugly bitch" from people who I doubt know what I look like. Every woman I know who calls herself a feminist, or is even just doing well, especially in a field in which men also contend, deals with some version of this, an underlying unease she evokes just by being a woman who doesn't devote every waking minute to making some man feel ten feet tall. Sure, you can brush it off, but that brushing off, over a lifetime, has psychic costs. And anyway, why should you have to?

Think of it this way: if all the castrating bitches voted for Satan's daughter, the ambitious lesbian robot, we might actually move the feminist revolution out of the parking lot where it has been sitting, low on gas and with major transmission problems, for the last decade and a half. Maybe the only way to defuse the immense fear so many Americans have of a woman assuming the quintessentially masculine mantle of the presidency, her delicate manicured index finger hovering over the nuclear button, is for them to experience it and get over it.

It's not just men who could use this lesson. If Hillary were president, women might eventually stop writing those dithery, snarky pieces about her clothes, her looks, her age, her marriage—pieces that manage simultaneously to acknowledge and to reinforce the double standard by which male and female politicians are judged. Face it, ladies. As long as women in positions of power are as rare as Florida panthers, their femaleness, with all it connotes, will be the lens through which people see them. We'll never be equal as long as "ambitious" is a dirty word when applied to women, as if male politicians are modest and self-effacing; as long as "serious" and "businesslike" read as "cold." It works the other way too: we'll never be equal as long as the president has to be the national daddy. But here's the paradox: The only way to make femaleness seem a natural quality in a leader is to have lots of women visibly running things. But as long as women are judged consciously and unconsciously by different, narrower, higher standards than men, that day will never come. The man will always get the benefit of the doubt.

And here's the problem: women are just as much—well, almost as much—a part of this double-standard system as men. Women don't like "cold," "ambitious," "angry" women either. That's why Hillary Clinton goes on the morning shows and girl-talks about dieting and clothes. For every woman who is excited by the prospect of voting for a woman in the presidential election—and let's not forget that Hillary, as of this writing, has a very large lead among women primary voters—there's another who feels it's somehow dishonorable to take gender into account. I know, because I used to be one of them. In 1992, when Elizabeth Holtzman and Geraldine Ferraro were running in the New York State Democratic senatorial primary, I managed to find reasons to reject both of them—Ferraro favored the death penalty, I think it was, and Holtzman had gone after Ferraro's husband's business dealings in a way that felt sensationalist to me. I cast my ballot for low-key, humane state attorney general Robert Abrams. I felt so lofty, so just, so rational! Abrams won, ran an invisible campaign, and lost to Alfonse D'Amato in November.

This time around I find myself thinking, what's wrong with putting the thumb on the scales just a little for a Democratic woman, all else being equal? Just to balance out the surprisingly large number of people (fortunately, mostly Republican) who tell pollsters they won't vote for a woman, however qualified? I would never support an anti-feminist woman, a right-wing woman, a Margaret Thatcher. But Hillary has stood up for women's rights for forty years. In the context of American politics, which to a European would seem to offer a spectrum that goes all the way from moderately conservative to insanely reactionary, she's a liberal. A lot of people think she's not electable, but the last time Democrats voted in the primaries for the candidate we thought would have the widest appeal in November, we got John Kerry. We think we can handicap Hillary's chances because she is the only candidate voters know well. But maybe Obama or Edwards would be the Robert Abrams of 2008, great on paper but lacking what it takes in real life. Maybe Obama will read too young or too black, or Edwards too slick or too one-note, to win over the undecided voters of Ohio or whatever demographic sliver will hold the key to victory.

Think of it this way. Hillary Clinton has been at the top of the list of Democratic candidates since the list began. She has a powerful campaign machine, which she began putting together more than a year before any of the other candidates got started on theirs. In fact, she's sewn up so much talent that, one British reporter told me, Al Gore would have a hard time finding good staffers if he decided to jump in. If she were a man there would be no doubt she'd win the primary, and the general election too. Oddly, it's right-wingers who seem most able to acknowledge, albeit unhappily, the strength of her position. "Looks like it will be President Hillary," one blogger wrote on the arch-conservative website humanevents.com. "Marxism here we come!"

Harridan. Virago. Whore. Ball-breaker. Stalinist. Hellcat. Evil Queen of Darkness.

As I write this in September, the New York primary is far in the future, so I have lots of time to see how the candidates and their campaigns play out. In the end I wouldn't choose Hillary *just* because she's a woman. The issues facing the country—Iraq, health care, education, poverty, civil liberties, Afghanistan, security, our burgeoning torture industry, reproductive rights, judicial nominations, and on and on—are too crucial to treat the election as primarily an opportunity for national gender reeducation.

Don't try my patience though, Rush, Chris, Peggy, and all you anonymous posters and bloggers out there. I'm only human. You might just push me over the edge.

4

POLITICAL ANIMALS

IS HILLARY A CAT PERSON OR A DOG PERSON?

BY SUSAN ORLEAN

Question: Hillary Clinton—cat person or dog person? Unlike many politicians whom you can peg immediately (doesn't Guiliani seem like the classic cat man? And can't you just picture John Edwards romping with a dog?) Hillary is hard to figure. If she's a dog person, exactly what sort of dog person is she? Would she favor a shih tzu? A Weimaraner? A rescued mutt? A Labradoodle? A pair of Nova Scotia duck-trolling retrievers? Perhaps she's a cat person. In that case, is she even electable? What if, unbeknownst to us, but in her heart of hearts, she's a bird person—and bold enough to put an end to the forty-year absence of pet birds in the White House? Is she, just maybe, not into animals *at all*? If she is elected president, might she usher in a brave new age, not just as the United States' first female head-of-state but also as the first president since the dour, mustachioed Chester Arthur to have no pet at all?

Hillary has, of course, had pets. When she lived in the White House, the Clinton family included the indolent, piebald Socks and the mud-colored Buddy. Socks the cat pre-dated Buddy the dog, which gives traction to the theory that Hillary just might have a predisposition for the kitties. If that's the case, you can't help but wonder whether Buddy, a choco-late Lab, might have been a bit of a beard, brought into the picture as a photo-op accessory (the fate of so many White House dogs!) but also, and more importantly, as a cover, an antidote to the wily, standoffish Socks. There is plenty of rea-son to suspect this was the case. Politicians who favor cats and don't throw in a dog to offset the assumptions one makes about cat lovers (namely that they, like their damn cats, are sneaky, aloof, self-licking, and not to be trusted around sleep-ing babies) have never done well in national politics. The last president who kept only cats was Martin Van Buren, and it just so happens that Van Buren's cats were Siberian tiger cubs. The Clintons, burdened during their White House years with Whitewaters and Monicas, might have needed a buddy like Buddy to soften their image. Nothing makes a powerful per-son seem more unthreatening than a waggy-tailed dog, and no dog seems more guileless than a Lab.

Socks may have dominated the Clintons' White House years, but he was not their first pet. Before Socks, there was the clownish blonde Zeke, a cocker spaniel with bug eyes and a mild overbite who was a gift from Hillary to Bill during his first term as governor of Arkansas. Zeke is largely unsung, al-though Hillary does mention him in the introduction of *Dear Socks, Dear Buddy,* the book of letters to the presidential pets that she edited. Zeke, according to Hillary, was "intrepid," "dogged," and so determined to wander that "he'd bite or dig his way through or around any barrier." There is a grudging respect in her description of him, and real admiration for Zeke's gusto and enterprise, but if you read the passage several times, it rings a little tinny. Isn't there a touch of disdain there—just a touch, though delicately delivered? Is there a slight absence of affection? Can you help but wonder whether Hillary, if she weren't so politic, would have complained that Zeke was annoyingly monomaniacal and so stupid that he might have gnawed through a live 220-volt line?

Zeke, like Buddy, was earmarked for Bill. Socks was acquired for Chelsea. Has Hillary ever had her very own pet? Her biographies and autobiography, describing her upbringing near Chicago, emphasize her devotion to the Brownies and the Girl Scouts, her success on the school safety patrol, her fondness for pinochle, her skill at basketball games like Horse and Pig, but there is no mention of any beloved childhood pet—no dog, no cat, no gerbil. Hillary's father, Hugh Rodham, was a tough, stingy, Republican grump, and he might well have been the kind of father who thought pets were frivolous and expensive, or perhaps too Democratic. Did Hillary yearn for an animal? She describes herself as a tomboy, which usually implies an ungirlish enthusiasm for picking your teeth with a jackknife and roughhousing with big golden retrievers. If the Rodhams had one—a golden retriever, that is, not a jackknife—the poor animal evidently made no mark, left no memory, and will not be inscribed in the book of presidential personal histories.

The question of politicians and their creatures is not an

idle one: the choice of a pet is deeply revealing of character, providing a sort of mammalian (and in a few cases, reptilian and avian) Myers-Briggs typological assessment that, in the case of a politician, might let you have an interior glimpse into what you are really getting for your vote. Woodrow Wilson had a ram. Andrew Johnson kept white mice. James Buchanan owned an elephant. Lyndon Johnson was a beagle-ear-puller. Richard Nixon, after riding to office on the reputation of his old spaniel Checkers, brought his weird, wild Irish setter, King Timahoe, to the White House. And it wasn't enough for the first Bush family to merely have their dog Millie; they had to have Millie multiply (flying in the face of the spay-and-neuter movement) so they could supply the next Bush administration with a Millie issue of its own. All these political animals were perfect fits, psychometrically speaking, with their masters.

So what about Hillary? Who is she? Animal-wise, she remains a cipher. At the moment, she's the only candidate to have inspired a constituency of four-legged friends (see "Pets for Hillary" at votehillary.org/CMS/Pets), but she also has been hounded for months by a furious antivivisectionist fruitarian critic who says he hates her for voting to bomb the animals of Iraq and for taking campaign money from chicken butcher Don Tyson. Darker still is Hillary's personal pet history. Zeke, that cute dumbbell, finally succeeded in his bid for freedom, busted out of the governor's mansion, dashed into the street, and was immediately hit by a car. After Zeke's untimely death, Hillary declared a ban on pets in the household because, as she writes in *Dear Socks, Dear Buddy*, "we felt so sad." As for Socks, he wore out his welcome once the Clintons left the White House and Chelsea left for college. Lucky for him, the president's loyal secretary, Betty Currie, gave him sanctuary in her Maryland home; otherwise, he might have been the first presidential cat to have been put up for adoption at the Washington ASPCA. Buddy made it out of Washington and up to Westchester, but not for long—he was killed by a car in Chappaqua in an incident blamed on Bill's inattention. Bad luck? Or is it bad husbandry? It's hard to know.

Recently, I was in Hillary's house in Washington for a fundraiser and poked around for signs of animal life. There is a beat-up doormat off the back porch that says, "Wipe your paws," but that's probably more for the amusement of guests than for the edification of pets. Framed family photos are scattered around the sitting room: Bill, Hillary, and Chelsea; Chelsea pirouetting; Bill and Hillary looking shockingly young; a handsome, candid shot of Bill hugging Hillary with one hand and holding a paper sack in the other. The inscription—scrawled in Sharpie, signed by Bill, reads, "Honey, even though you left me holding the bag, I love you." Beside that, in a dark frame, was a fuzzily focused picture of a chocolate Lab—Buddy—all by himself, striding purposefully through something that might be sand or snow. That was it. The house was elegant, book-lined, chintz-upholstered, and clean as a whistle. There was a noticeable absence of both hairballs and chew toys, which makes me think that Hillary is now entirely neutral when it comes to animals—she has probably sworn off pets for good.

5

BOYS AND GIRLS

HAS THE MOMENT FOR FEMININE ROLE MODELS PASSED?

BY LORRIE MOORE

When most of us first laid eyes on Hillary Clinton, it was on *60 Minutes*, right after the 1992 Super Bowl and she was sitting by her man saying she was not one of those women who would stand by him. Nor did she make cookies. That part I liked—I thought I heard the thunderous cheer of nonbaking working women across the country, but apparently I was wrong, and soon we the people were favored with Mrs. Clinton's very own contrite cookie recipe, which I have not yet had time to try.

But on that Sunday night fifteen years ago my ears perked up and my eyes swooped down: who was this blonde with the headband and the regionally unplaceable accent? Where was she from—and were headbands *back*? Why did she so blithely distinguish herself from Tammy Wynette, who knew better than to stand by any number of men? (And whoever Gennifer Flowers was, should we care? The name alone seemed satirical, even for a stage name. It was the mysteriously banal Hillary I was watching. Mrs. Clinton was clearly a personal and political partner of Bill's—was the electorate ready for this?— but when the interviewer questioned them as to the nature of their marriage, referring to it as an arrangement, it was not Mrs. Clinton but Mister who took the bait, tossed it to the floor, and said a little heatedly, "Wait a minute: This is not an arrangement. This is a marriage." And I'm sure to him it was, and is, still, in its way. But Hillary's seeming hesitation on this matter—we'd just watched all kinds of helmeted people running great lengths to catch all kinds of passes—spoke the complex volumes that every woman knows: marriage *is* an arrangement.

That anyone should be faulted for using his or her connections to run for public office seems a little laughable, since that is what politics has ever involved: "friends." And as for Hillary's fleet opportunism—advance planning we were slow to become aware of, plus those Sunday school virtues of patience and hard work that may have made her moves seem less eager and grasping to her than they did to others—well, many have been there before. In 1966 Robert Kennedy made the same mad, carpetbagging dash to New York to pick up that very same Senate seat in a state he was not really from, clearly with an eye to the 1968 presidential campaign. And there was even a smidgeon of poaching in Wisconsin's beloved William Proxmire, who was really, like Hillary, a Chicagoan, and lived very little in Wisconsin, even while senator. And if Hillary, despite her great media exposure, or perhaps because of it, seems a little unknowable—Norman Mailer once said of Jackie Onassis, after he stood next to her in a blinding, soul-blitzing blaze of camera bulbs, that in that kind of glare Onassis could not possibly know who she was—we still feel friendly enough as a nation to refer to Clinton by her first name. I don't believe it is disrespect or sexism that does that. I think it's jazz. Albeit, journalism jazz.

As has been noted by many, Hillary's public personality is too often in pragmatic retreat, overmanaged, increasingly botoxed and schoolmarmish, a frozen text. But when her face cracks open in a big laugh—or when she refers to Dick Armey as the henchman of Dr. Kevorkian (this can be glimpsed in Michael Moore's documentary *Sicko*)—there is no not liking her. I think even Dick Armey liked her there for a second. But she seems to many eyes to be a person in great flux. It took a year or two of college to shake off her Republican affiliations—why so long? No teenager I've ever heard of, not one so destined, has taken that long, no matter who her parents were. (Note Patti Davis.) She seems to hold many contradictions simultaneously—is this the sign of first-rate intelligence, as Fitzgerald would have us believe, or of an inner saint and an inner thug playing out their options behind a chameleon's façade? Throughout her life she appears to have continued paradoxically in both a searching way and a highly planned way. The marriage is part of the plan (with Chelsea still young, leaving the marriage would not have been any more feminist than staying in it, nor would it have bestowed upon Hillary any special dignity—would she then be, god forbid, *dating*?—divorce and pride are often actually debilitating things). The shifting hairstyles, the unstable voice, the hide-and-seek persona, the cookie rebuke and insult to Wynette, the national health-care blueprints begun then botched, are all fallouts of the search.

Because we have currently in office, in the words of Jane Smiley, not just the worst president in American history but the worst *possible* president, we still should not be tempted toward casting a rosy haze over the Clinton years in the White House; they were not years of great accomplishment. Baghdad was strafed and embargoed; Waco was gassed and burned; Oklahoma City was bombed: in all these, children were appallingly killed and Bill Clinton was the president. While polar ice caps began to melt, Al Gore was left to do god-knows-what in the Blair House, only to regale us years later in cineplexes with the catastrophic consequences of those melting caps. NAFTA and GATT were signed, the National Health Plan went nowhere, and by the second term's end the White House's hope to get things done had been hijacked by Ken Starr and the party dress he had confiscated from the closet of someone named Monica Lewinsky. Sure, Bill Clinton was our first American president to feel comfortable around black people—and this meant a lot to both black and white folks looking on. But it was in lieu of policy: JFK, LBJ, and Jimmy Carter were more uncomfortable but they did better.

So here comes another Clinton, with the same Clintonesque compromises, centrist sway, and self-serving parsing of the language (oh, to listen to her on her Iraq vote in the Senate). For years we watched her looking like a nerdier Tina Brown, who in turn looked like a nerdier Princess Di (the 1990s were confusing!), but now confidence has made her more beautiful than either. Strangely, there is this to be said about Hillary Clinton: I often hear her defended better by men than by women. Men see her as a victim of sexism. Women less so. Women see her as a beneficiary of celebrity. They see her not as one of them. But how could she be one of them? Her life has resembled no one's at all; women vibe her freakishness. And women are merciless with one another: they feel men are fooled by women but they, women, can see clear through one another. But they also want sisterhood, and sisterhood is hard with Hillary. Seeing clear through is not possible. She *is* a freak.

But anyone who runs for president is a freak—presidential candidates do not represent any ordinary person at all, though they always pretend otherwise. The only U.S. senator I know personally—a left-wing political maverick who votes outside the usual fray, but an ordinary Joe in many ways (last I spoke to him he was down with the proles, looking for his own suitcase at the LaGuardia baggage carousel)—has wanted to be president since he was 9 years old. What kind of people have wanted to be president since they were 9 years old? These are the kind of people all the candidates are—and Hillary is no exception. Her ambition is no different in nature from anyone else's: it's naked and unattractive and willing to make a few deals in order to have some power, er, rather, ability to serve. That's just how it is with all of them.

But does her being a woman make her a special case? History aside, does her gender confer special status and meaning on her candidacy? In my opinion it is a little late in the day to become sentimental about a woman running for president. The cultural moment for feminine role models may have passed. The children who are suffering in this country, who are having trouble in school, and for whom the murder and suicide rates are high—as well as the crime rate and the economic dropout rate—are boys, especially boys of color, for whom the whole educational system, starting in kindergarten, often feels a form of exile, a system designed by and for white girls. In the progressive midwestern city where I live, the high school dropout rate for these alienated and written-off boys is close to 70 percent. Some of them are middle class, but most are just hanging on, their families torn apart by the splintering forces that have crashed down on them, especially the harsh drug laws that have sent far too many people to prison. Why, for instance, has no one in the Democratic Party campaigned to have convicts who have served their time reenfranchised and made full citizens again? Their disenfranchisement is a continuing strike against these men, their families, and their children. The glitzy ads—spoofs!—that donation money is already being spent on are silly and vaguely sickening. Let us give this money not to the circus but to the kids.

Perfect historical timing has always been something of a magic trick, finite and swift. The train moves out of the station. The time for inspiring middle-class white girls, the group Hillary Clinton represents, was long ago. Such girls have managed on their own (given that in this gilded economy only the rich are doing very well). They have their teachers and many other professionals to admire, as well as a new Speaker of the House and a couple of Supreme Court justices and world leaders, from Queen Elizabeth I to Golda Meir. Whether this offsets Paris Hilton, well, perhaps Ms. Hilton will offset herself. The landscape is not bare. Boys, sadly, are faring worse, and Hillary Clinton's gender does not rescue society from that—only her policies will. Barack Obama may have less trouble inspiring our children; he is smart and human and scarred (so many of the current Democratic candidates have suffered terrible family tragedies, and he is the youngest) and this lends him the slight sadness that spring-feeds warmth and compassion. He also grew up as a black man in America, whether he or anyone else fully appreciates this, and he does not need (as he has done) to invoke his grandfather's life in colonial Kenya to prove or authenticate his understanding of race in this country. His sturdiness is equal to Clinton's, but he brings a newer and so more hopeful and exciting face, domestically and abroad.

That may not matter: we need more than something as superficial as a face—male or female, black or white, Clinton or not. The parents of African American boys, the leaders of the world, the workers of this country, all want real and difficult and heartfelt policymaking. The melting ice caps threaten to drown millions; the debacles in Iraq and Afghanistan bungle on; health care is like the Borscht Belt joke about the food being bad and the portions too small. After Obama has courted the youthful fringe and Clinton the center, and they set their squabbling aside (and I hope this can be done), I look forward to them coming together, reuniting the party and becoming the ticket. She may not be a thrilling person to vote for. Still, it would be a thrill to see her win.

6

THE BALLAD OF BILL AND HILL

CONFUSING THE ROMANCE OF POWER WITH THE POWER OF ROMANCE

BY DAPHNE MERKIN

If other people are an enigma, other people's marriages are a conundrum wrapped in a riddle nestling inside a puzzle encased within impenetrable walls. We conjecture, we imagine, we project, and we deduce, dancing rings around an essential lack of hard information. Perhaps the institution's obdurate refusal to be stripped of its cover, unlike that naked Original Couple who led each other down the garden path, in full view of any Tom, Dick, or Harry who happens to pick up the book of Genesis, is marriage's way of preserving itself. In any case, for all our poking around for evidentiary signs and clues, other people's marriages hover tantalizingly out of view, casting a pale light from time to time, like the moon passing behind a cloud, but mostly leaving us in the dark—and the parties involved to their sovereign privacy. It may well be that in this age of incessant media exposure we no longer consider married couples entitled to keep their secrets, but keep them they do.

The marriage of Bill and Hillary Rodham Clinton has been a subject of fascination to friends and to strangers ever since they took their vows on October 11, 1975, in Fayetteville, Arkansas. "Probably more has been written or said about our marriage," Bill Clinton observed in his memoir *My Life*, "than any other in America." In May of 2006, an article appeared on the front page of the *New York Times* in which a reporter attempted to lift the veil (make that veils, seven of them) from the Clintons' conjugal life by interviewing fifty people chosen at random. Among other things, the article laboriously parsed the couple's hectic schedules as though they were entries in an accountant's ledger, revealing that, since the start of 2005, the Clintons had been together, on average, two weeks a month; had at least brushed by each other on twenty-four out of thirty-one days; and had spent fifty-one of the last seventy-three weekends together. Go figure. Still, if you bother to do the math, what do these numbers really tell us? That the Clintons sedulously avoid each other? Or that they try to be with each other as often as their busy lives allow?

Books and articles continue to appear, rumors abound (that Hillary is a lesbian or, at the very least, is not bothered by his unremitting flings; that he asked for a divorce in 1983, which she refused to give him), and questions persist: Are they in some kind of love? Were they ever in love? Do they even like each other in spite of what must surely be smoldering resentments on both sides? Or do they represent a very contemporary arrangement, one that is either wholly unsentimental or overtly cynical, in which they are bound together by a ruthless, jointly conceived drive to reach the pinnacle of American politics? (The fact that they may also be bound together for that most basic and conservative of reasons—for the sake of their daughter, Chelsea, to whom they appear to be mutually devoted—rarely enters the conversation.)

Or, then again, perhaps theirs is a symbiotic John-and-Yoko union, one which might translate into a version of the famous record cover in which the couple were shown intertwined with each other in a primordial embrace. The sort of union that includes an inversion of traditional gender roles and no small amount of psychological sadomasochism, an almost reflexive dynamic in which each needs and facilitates the other by trading places as victim and victimizer. In this version of things, he is the yin to her yang, a romantic who thrives on adoration while she keeps her eye steadfastly on the ball. She does the strategizing behind the scenes and he does the emoting that brings in the crowds. His randy behavior is an important part of the psychological package, eliciting first her fury, then his penitence, and finally her forgiveness, cementing them together in a frayed yet workable form of forever-afterness.

Dick Morris, the astute and unprincipled political guru who was close to Bill and Hillary over a period of twenty years before he turned against them, has a take on the Clintons' marriage that confirms this construction. His analysis, which Carl Bernstein quotes in his biography of Hillary Clinton, *A Woman in Charge*, draws heavily on the vocabulary of co-dependency, with Bill radiating a wounded-bird masculinity while Hillary responds with an "everything is under control" femininity. "I believe," Morris observes, "that it's a relationship in which she is . . . addicted to him . . . that it is a relationship based on mutual enabling. Because she likes what happens when she rescues him . . . I think that he sometimes resents her and shakes under her domination. Sometimes he welcomes her and needs her, because he requires her rescuing."

Both Clintons like to tell the story of how they met, at Yale Law School: Bill was struck with Hillary's aura of poise and strength, and she was immediately drawn by his animal magnetism. In her memoir, *Living History*, Hillary calls her husband "a force of nature," and she seems to have been dazzled by every detail about him, right down to the look of his hands: "His wrists are narrow," she writes, as though coming upon one of the Seven Wonders of the World, "and his fingers tapered and deft, like those of a pianist or surgeon." (The only woman who appears to have gripped Bill's romantic imagination in a similarly consuming way was Virginia Clinton, his "beautiful, high-spirited" mother. A brief but revealing passage in *My Life* describing Bill sitting on the bathroom floor as a little boy, watching his widowed mother prettify herself before she went off to her job as a nurse-anesthetist, says more about the shared narcissism that characterized his connection with his mother—and would later sustain his relationship with Hillary—than any shrink's assessment: "It took quite a while, partly because she had no eyebrows. She often joked that she wished she had big bushy ones that needed plucking, like those of Akim Tamiroff, a famous character actor of that time. Instead, she drew her eyebrows on with a cosmetic pencil. Then she put on her makeup and her lipstick, usually a bright red shade that matched her nail polish."

Sparks or no sparks, Bill had to propose to Hillary several times before she accepted. According to Jeff Gerth and Don Van Natta Jr., the authors of *Her Way*, which is probably the least partisan book written about Hillary—exuding neither the animus of Michael Tomasky's *Hillary's Turn* nor the cozying-up approach of Bernstein's biography—she needed time not only to figure out whether their marriage would endure (Hillary's mother, Dorothy, had borne the ill effects of her own parents' divorce) but also to decide whether she was willing to bet the odds on Bill's political future. The calculation would have to include trading in her own ambitions and her Chicago-bred, Wellesley-nurtured sophistication for his impoverished and hillbilly home state, known primarily for the size of its watermelons and barbecue shacks. In what was either an instance of the heart wanting what it wants or, according to a less benign view of their pairing, the first step in what has been described by several writers (including Bernstein) as their "twenty-year pact" to reach the White House, to the amazement of many of her friends, Hillary decided to cast her lot in with Bill.

The two were married in the front room of a tiny redbrick house with a screened-in front porch and no air-conditioning that Bill had made a down payment on ($3,000 toward a purchase price of $20,000) only weeks before. Bill was 29 and had shaved off the beard and long hair he had worn at Yale; Hillary was 28 and frail-looking in a steely sort of way. A Methodist minister presided and Hillary, who had not wanted an engagement ring, wore a Victorian-style dress she had bought the night before with her mother at a local department store. The couple's honeymoon was postponed for two months, and when they finally took off for Acapulco, Hillary's parents and two brothers—as well as one of her brother's girlfriends—bunked with them in the presidential suite. According to *My Life*, they all had a grand old time of it, "playing pinochle" and "swapping stories." Still, for a man who thought that he'd never get married, Bill must have found this step a bit overwhelming—which perhaps helps explain why his honeymoon reading included the somewhat incongruous choice of Ernest Becker's *The Denial of Death*. (This is one of the books that Woody Allen gives Diane Keaton in *Annie Hall*, as if it might lead to a mutual understanding of the existential abyss that beckons beyond all our earthly arrangements.)

ONE REASON, OF course, that the Clinton marriage has provoked so much speculation is because the specter—or is it the mystique?—of ongoing sexual misbehavior has colored our perceptions of it from the start. He is an intractable philanderer; she is constantly mortified, but resilient. The disparity between their physical appearances—he is extraordinarily handsome while she is passably attractive—has further fueled the discourse, leading some people to wonder why he hasn't left her for a more beautiful and adoring woman, and leaving others *not* to wonder at the fact that she has stuck with him despite serial and very public humiliations. In a 2003 interview in *Time* magazine, Hillary disclosed that she and Bill had started marriage counseling in August 1998, when the Lewinsky scandal was at full throttle. She warded off specific inquiries into how much she had known about the earlier allegations about Gennifer Flowers and Paula Jones, as well as the extent of her husband's lying, by musing, instead, on the bewildering thing that is matrimony: "I don't try to make any judgments about any other people's marriage, because it's a mystery—why two people are attracted to each other, why they love each other, why they marry, why they stay married."

So many mysteries, so little time. Although it's hard to believe that a woman as strong-minded and self-righteous as Hillary doesn't have firm opinions about how other people conduct their marriages, there is something shrewd about her casting the revelations that roiled her own marriage as standard issue rather than as a deviation from the norm. After you sign on, she appears to be saying, anything can happen; in the instance of her marriage, it turned out to be a roller coaster that she has chosen not to get off of. (The interview ended with Hillary staunchly declaring that she had "no intention of running for President" in 2008, which suggests that political dissembling and emotional dissembling often go hand in hand.)

In some ways, the Clintons' backstory recalls that most basic of boy-girl narratives, the subtext of which hasn't changed in the thirty years since the couple exchanged old family rings in a wedding that took place without fanfare in front of twenty friends. They are suspended in the alembic of their relationship as it first was: he is still the best-looking guy on campus, with charisma to spare and a seductive way with the opposite sex, and she remains the straight-arrow girl with coke-bottle glasses and frizzy hair who managed to nab him for herself. One wonders if Hillary can possibly be as inured to her husband's affairs as she seems to be. Diane Blair was Hillary's closest friend before she died of lung cancer in 2000 and played a key role in the group of mostly female aides whose job during the 1992 campaign was to defend Bill from attacks on his record as governor of Arkansas, his avoidance of the draft, and his alleged dalliances; when asked by a colleague, during the Gennifer Flowers dustup, why Hillary always stood by her man, Blair's answer was as crisp as a package of saltines: "Hillary knew what she was getting into when she married him," she said.

The truth is, Hillary has never had much use for the sort of introspection that might make her confront whatever suffering she has sustained because of Bill's wandering eye. And notwithstanding the vogue for get-it-all-out-in-the-open, intimacy-obsessed, problem-solving notions of marriage, there are things to be said for this kind of anti-psychological approach. It helps you keep your focus on the road ahead, for one. Bill had always strayed, well before Gennifer Flowers, and Hillary had always waited it out. The fact that his other women are so far from Hillary's "type" adds a poignant touch to the situation, underscoring the irresolvable mind-body, madonna-whore complex that plagues many men.

What is clear is that his intellect goes one way and his zipper another, allowing him to implement the vision of America that informs their "twenty-year pact," while dooming him to furtive gropes and official inquiries into the "distinguishing characteristics" of his penis. The entrance of Monica Lewinsky, bearing the gift of her thong, upped the ante and gave the Republicans who had been on Clinton's trail the ammunition to initiate impeachment proceedings. But Lewinsky did not appear to be a difference in kind so much as in effect. Although many marriages involve infidelity (usually on the husband's part) and tolerance of bad behavior (usually on the wife's part), there is something mega-sized about the fissures in the Clintons' façade, creating holes big enough for everyone to try and peek through.

One could speculate forever about the mechanisms that keep the Clintons a going concern and still come up empty-handed. You can throw whatever darts you like at them and still not get anywhere near the bull's-eye of what makes them a team. In the end, their marriage strikes me as being infinitely more complex than we give it credit for. Echt feminists, who tend to see strength where others might see compromise, have rushed to claim the marriage for their cause. Gloria Steinem once referred to it as an example of "the family as democracy"; and Naomi Wolf, a next-generation Steinemite, has called it an "equal-partnership family." Each to her own vision of a brave new world. As Dick Morris once told the president, "Look, you and I know the reality of your marriage. Your strengths feed on each other. But people don't get it. They think either she's wearing the pants or you're wearing the pants." (Still, although their marriage may not be as simple an inversion as that, one gets the unmistakable sense that, of the two of them, Hillary comes equipped with Republican balls and a grasp of political expediency while Bill comes equipped with the Democrats' sense of empathy and belief in the rightness of the common cause.)

I would suggest, finally, that one of the ways we stumble in trying to figure out this couple is to confuse the romance of power with the power of romance. Power, as we well know, is an aphrodisiac. The audacity of ambition—which is what the Clintons expertly juggle between them, like two circus performers—must bring an erotic charge, a rush of mutual pride and admiration that is hard to beat. Alone, one or the other might falter or get diverted on the path to political ascension; together, as their 1992 campaign song energetically assured us, they never stop thinking about tomorrow.

7

MONARCHY IN THE MAKING

WHY ELECTING A FORMER FIRST LADY IS AN ANTI-FEMINIST STORY

BY LIONEL SHRIVER

Stories are not merely the province of fiction. History is made of them, and they matter. The *story* of how the first woman is elected president of the United States will last far beyond the tenure of her administration, and will have repercussions for the reputation of her gender nationwide long after she has stepped down. It is on the basis of story that I would be sorry to see Hillary Clinton, wife of a popular previous president, become the first female leader of my country.

That story is anti-feminist: former First Lady is elected on her husband's coattails. Having the same insidiously undermining effect as affirmative action, ascent to the White House by marriage casts doubt on whether a woman could make it without help. Hillary's qualification for the office is massively premised on her marriage to Bill, who, in an inversion of the old saw "behind every great man stands a woman," would implicitly fill the role of the great man behind every girl.

For kindred reasons, I opposed George W. Bush's candidacy for the same office. There were other issues, of course—his pious religiosity, his inarticulacy, his weak mastery of foreign affairs. I would never claim to have foreseen the catastrophe of Iraq, so at the inception of his campaign what put me off was the prospective story: son of previous president steps into his father's shoes. That is the story of a monarchy.

Having lived outside the United States for two decades, I may be somewhat more sensitive than my compatriots in-country to how our nation's affairs appear from afar. The nepotistic election of George II—even sounds like a king, doesn't he?—made the United States look hypocritical. In office, George W. has thrown around the word "democracy" almost as often as the word "evil." But the moral of his own story is that to make it politically in the United States you have to have connections. That the election of the American president is one more backroom deal, the bunting and baby-kissing being all for show. That in the United States, like everywhere else, power is concentrated in a few hands and meted out sparingly to cronies and relatives.

Of course, American political families go back to John Adams, the second American president, whose son, John Quincy, became the sixth. More recently, John F. Kennedy spawned a dynasty to which few Americans have objected. For Democrats, it is tempting to see his brother Bobby's subsequent, tragic candidacy for president and the distinguished senatorial career of his brother Edward as a benevolent matter—merely the consequences of an especially talented family's willingness to serve the public good. Yet the fact that these politicians have roughly lined up with my own party leanings makes the story of the Kennedy dynasty no more appealing. Bobby and Edward weren't only gifted; they had connections.

Similarly, the United States has a sporadic tradition of elevating politicians' spouses to public office, or at least of considering them as more viable candidates than they might be regarded had they not married into the political fold. Thus, Elizabeth Dole followed her husband Bob in pursuing the Republican presidential nomination and is currently serving as a senator for North Carolina. (In her defense, Elizabeth Dole had a long, illustrious political career underway before she ever married Bob, serving in both the Johnson and Nixon administrations, and under her own steam, eventually becoming secretary of labor under George H. W. Bush.)

Mel Carnahan, a Democrat running against John Ashcroft for the U.S. Senate in Missouri in 2000, died on the campaign trail in a plane crash. The ballots had already been printed. His widow, Jean, unofficially replaced him, while the campaign continued to employ her husband's original slogan, now taking on a morbid cast: "I'm Still with Mel." After the voters elected a dead man, as promised, the governor appointed Jean to the office. Jean's appointment and the votes that the deceased candidate garnered were sentimental and not based on her qualifications for the Senate. Uncomfortably, I cannot think of a prominent male politician who owes his success to having been married to a political wife who was high-flying—figuratively or literally.

Does being married to a president qualify you for the job yourself? Without a doubt, Hillary would have a better idea than most of just what the job entails. But patience through dreary state dinners, renovation of the Blue Room, and even what-will-we-do-about-Monica pillow talk is a far cry from crafting government policy; Hillary's health-care initiative, her primary foray into policymaking during her husband's administration, was stillborn. What other professions confer a husband's professional expertise on a spouse? If a woman came at me with a scalpel, reassuring me that she knew what she was doing because her husband was a surgeon, I'd flee for the exit in my hospital gown.

It is highly unlikely that Hillary Clinton would ever have become a credible candidate for the presidency absent her marriage to Bill. If nothing else, she is—I was about to write "charmless," but I have had just enough experience of being on the receiving end of the blithe, brutal personal insults hurled at public figures to have grown leery of hurling them myself. So I will rephrase: Hillary Clinton is not a natural public speaker. Arguably, the same qualities that make Hillary so wooden before an audience are the ones that make her appealing as a person—humility, an inclination to privacy, a constitutional reluctance to bask in the limelight. (By contrast, Bill swells in front of a crowd like a tick.) Indeed, about the only thing I ever liked about George W., also no political natural—and that is an understatement—was the initial discomfort he displayed with having to perform and be the center of attention. During his first painful months in office, Bush's unease with public speaking was almost endearing. He was bad at it, but at least in the olden days he seemed to realize that he was bad at it.

Yet by 2009, we will have had eight unrelenting years of a president who cannot utter a sentence without using the same word five times, who has single-handedly bolstered the novelty book trade by providing a festival of verbal gaffes to purchase at the Barnes & Noble checkout. I yearn to watch an American president giving a press conference on television who does not drive me, cringing, behind the couch. Although Hillary would probably not prove another outright embarrassment, her dry, robotic oratory is anything but inspirational and would surely have alone knocked her out of the running for the Democratic presidential nomination were she not allied in the public mind with a certain someone.

While Hillary has finally acquired her own experience in political office by serving two terms in the Senate, she also owes that seat to the fact that she is Bill Clinton's wife. Having just moved to Chappaqua the year before, a long-term resident of Arkansas born in Illinois would not otherwise have been elected as a representative of New York State. In Arkansas, for twelve years she was known primarily as the wife of the charismatic governor. Granted, she practiced family law, at which she was skilled—but how many other women in the country during those decades were competent lawyers? Hillary is clearly an intelligent, self-possessed professional, but nothing about her biography leaps out to distinguish her from droves of smart, levelheaded peers who didn't happen to marry a president-to-be.

The "Iron Ladies" Golda Meir and Margaret Thatcher both gratifyingly belied the assumption that female leaders would be flighty, emotional, or soft; during their tenure, no one made jokes about how either might just press "the button" during the wrong time of the month. Although unjustly beset by gossip about her hair and occasionally playing to female stereotypes by publishing her recipe for chocolate-chip cookies, stylistically Hillary is at least firm. If not as formidable in bearing, like Meir and Thatcher before her she doesn't appeal to the electorate by being sexually beguiling. (Ségolène Royal tried the more flirtatious route in her failed bid for the French presidency in 2007.) Yet neither Meir nor Thatcher owed their rise in the political ranks to marriage. Rather, the UK's imperious prime minister introduced to the British lexicon a new genre of reduced masculinity: "a Dennis Thatcher"—the pallid, humble male helpmate in the shadows. It is hard to imagine Bill Clinton as a Dennis Thatcher.

To the contrary, all too many voters in the 2008 primaries will be casting their votes not for Hillary but for Bill—who has made no secret of his exasperation that the Constitution prevents him from running for a third term, and even hinted with no little chutzpah when he stepped down that the Constitution might be changed to allow him to do so. Now he seems to have devised a route into the White House through the back door. In the event that Hillary were sworn in with her hand on a Bible in 2009, plenty of voters would be relieved to catch a wink from the charmer at her elbow as to who's *really* going to be running the show. That Bill would be as much at the tiller as his wife in an effective "co-presidency" is a major thesis of Carl Bernstein's 2007 biography of Hillary Clinton—titled, ironically, *A Woman in Charge.*

On the release of that biography, the headline in London's *Daily Telegraph* read, "Two for the Price of One, If Hillary Wins." The subhead said it all: "The Clintons will operate a joint presidency, with Bill using his charisma and experience."

If Bill will be using his charisma and experience, what will Hillary be using? The title. Etymologically, that's where the word "titular" comes from, whose first definition in Webster's is "1) a: existing in title only: NOMINAL b: having the title and usu. the honors belonging to an office or dignity without the duties, functions, or responsibilities." Do we really want the first female American president to be, in the eyes of the American populace and the world, a titular head of state?

The embarrassing truth that Bill more than Hillary is the electoral draw, and perhaps even the real candidate—the unofficial candidate, like Jean Carnahan—has been evidenced in the early days of her campaign. In Iowa in July 2007, with Barack Obama breathing down her neck, Hillary finally brought out her husband like a secret weapon. Clearly making enormous efforts not to overshadow her, like a friendly Great Dane on a very short leash, Bill constrained his humble introduction to nine minutes and thereafter literally kept his head down—seating himself on a three-legged stool about ten inches high. The British journalist Toby Harnden described the scene: "Whereas Mr. Clinton exuded an almost instinctive empathy with the crowd of several thousand, his wife's chopping hand gestures and flat, almost hectoring style of delivery led to a rising murmur from the fringes as people's attention drifted and they began talking." After their Des Moines rally, while both Clintons signed T-shirts and baseballs, it was Bill who collected the noticeably larger crowd. For Hillary, her husband's conspicuously greater popularity is humiliating, his little-ole-me routine on stage perhaps even more so.

I would never vote for a presidential candidate solely because she is female, any more than I would vote for Barack Obama solely because he is black. A woman is inherently no more in line with my political values than a man. Yet were she to roughly represent my views, I would relish seeing a woman become president of the United States. To date, American women have made inroads into the once exclusively male domain of political power. At this writing, nine governors are women, as are sixteen senators, and seventy members of the House, to which Nancy Pelosi's election as speaker was especially symbolic. But women are still a far cry from constituting a fully proportionate half of those elected to public office. A female president would betoken gender equality in the United States at its highest levels, and could inspire more qualified women to enter politics, as it would also inspire more voters to elect them.

When I was coming of age in 1972, the black congresswoman Shirley Chisholm didn't have a prayer as a presidential candidate. A vote for Chisholm in the Democratic primaries was a point scored for gender and racial equality, but otherwise it was a vote thrown away. These days, a woman becoming president is a credible turn of events, and at least no one is arguing that a vote for Hillary is a waste of electoral capital that could otherwise be spent on a candidate who might actually be nominated.

Back in 1999, I objected to George W. Bush's candidacy on primarily narrative grounds—because handing down the presidency like a crown from father to son made the nation look two-faced and corrupt. I have subsequently had to concede that in those days my priorities were skewed. The story of electoral nepotism should have taken a backseat to the even larger story—what our newly coronated king would do—for the invasion of Iraq may well prove the most egregious foreign policy mistake in the history of the country. Ergo, if Hillary Clinton is nominated as the Democratic Party's candidate for 2008, and the Republican who opposes her seems comparatively less trustworthy as a steward of my country's interests, I will have no choice but to bite the bullet and vote for her. But it will hurt my teeth.

We are not in full control of our own narratives. My story, such as it is, has been heavily colored in the eyes of others by the facts that I was born American and female. I do not recall ticking the box for either on the way out of the birth canal. Equally not the master of her tale, Hillary Clinton married a man who later became president of the United States. Surely she is neither scheming nor prescient enough to have anticipated that advent with an eye to snagging the office for herself. By dint of accident, luck, and good or bad taste in men, depending on your perspective, she is now poised to become the first truly viable female candidate for the U.S. presidency. If I were Hillary Clinton, I would doubtless take advantage of circumstance, bow to the wishes of the many supporters in my party who saw my candidacy as a route to getting both a woman and a Democrat into the White House, and run for president. But I am not Hillary Clinton. I am watching her. I am following her story. For a feminist, that story is foul.

8

ALL HAIL BETTY BOOP

Remember when the idea of a female chief executive was a gender-bending gag?

BY REBECCA MEAD

The last time a woman was elected to the White House it was the summer of 1964. The successful candidate was a former judge and mother of two, as well as being the wife of an intermittently supportive businessman who never expected that his spouse, when she stood for office, would be embraced by 40 million American women eager to see one of their own in, rather than adjacent to, the seat of power.

This, at least, was the premise of *Kisses for My President*, a comedy starring Polly Bergen as Leslie McCloud, the attractive and youthful president, and Fred MacMurray as her beleaguered, bothersome husband. In *Kisses for My President*, the nation's first female leader is depicted as worrisomely dedicated to affairs of state rather than those of the heart and hearth. "I treasure these moments we have together, even if you don't know I'm here," MacMurray says over breakfast one morning to a chief executive immersed in her papers. The lot of the overlooked First Lady—along with that of the overlooked American housewife—was transformed by *Kisses for My President* into a gender-bending gag.

The movie is one of surprisingly few efforts on the part of the entertainment industry to imagine a United States in which the top political office is held by a woman. Betty Boop ran for president in 1932 in a cartoon that was really an expression of support for President Roosevelt. On television, we have had Patty Duke as president in *Hail to the Chief*, a short-lived sitcom about a female first executive with a philandering husband that was broadcast in 1985. Then in 2005, Geena Davis appeared in the ABC television series *Commander in Chief* as President Mackenzie Allen. A political Independent, president of a university, and ratings winner among soccer moms, Allen has been swept to power not by the American electorate but by an act of God, having been adopted as a vice presidential running mate by a Republican candidate who has the unfortunate political luck of suffering a fatal stroke halfway through his first term. The show was a drama: *The West Wing* as commissioned by Lifetime, with an empowered if vulnerable Davis boldly ordering military actions against rogue nations while vainly struggling to monitor her younger daughter's sugar intake in the face of the White House's twenty-four-hour room service. (What Allen's creators ultimately hoped for her is lost to history: the show was canceled in its second season.)

In *Kisses for My President*, however, the prospect of a female chief executive was played for broad humor, with the prospect of a First Gentleman in particular an opportunity for bald ribaldry. Thirty-five years after its release the movie is decidedly lacking in hilarity—not, it should be pointed out, that it was splitting many sides even in the 1960s. (The critic Bosley Crowther complained in the *New York Times* that it "proceeds from one corny contrivance to another." Its sole Oscar was for costume design.) Then again, the movie was released at a particularly unfunny moment in American political life. President Kennedy had been assassinated only months earlier, a tragedy that went unmentioned in *Kisses for My President*. An intractable war was under way. But the turbulence in American life that the screenwriter was most concerned to lampoon was the burgeoning feminist movement. A year earlier, in her bestselling book *The Feminine Mystique*, Betty Friedan had characterized a pervasive malaise among understimulated American women as "the problem that has no name." In *Kisses for My President*, the problem has a name: it is Leslie McCloud, the suspiciously androgynous-sounding Madame President.

The suggestion of the movie is that in order to be a success as a president, a woman must fail as a wife and mother. McCloud's 10-year-old son uses his police protection as a cover for inciting fistfights with his classmates, while the teenaged "first daughter" is apparently nurturing an incipient eating disorder and sneaks out of the White House at night to hook up with a hot-rodding hoodlum, the only boy in her school not intimidated by her Secret Service detail. (This theme is revisited in *Commander in Chief*, in which Allen's Republican-leaning teenaged daughter, Rebecca, grumps about the White House, adolescent rebellion butting up against affairs of state. "You be John-John and I'll be Patti Davis," Rebecca tells her handsome twin brother, Horace, Mom's biggest supporter.)

In *Kisses for My President*, the most pitiful member of the McCloud clan, though, is Thad McCloud, the nation's first "First Gentleman," who suddenly finds himself without a job but with a social secretary dedicated to scheduling his luncheons with senators' wives and his attendance at flower shows. When offered the opportunity to head the men's division of a cosmetics company founded by an old flame, he devises slogans for putative new fragrances such as "If you want to feel like a man, smell like a man," and fails to recognize that he has been hired less for his marketing abilities than for his marketability as a national novelty nonpareil. (There are, it turns out, a limited number of jokes that can be made about the emasculation inevitably experienced by a husband when he becomes First Gentleman: in *Commander in Chief*, Mackenzie Allen's husband, Rod Calloway, who has hitherto been his wife's chief of staff, is also ushered into an office decorated in pink and is attended by a social secretary who keeps trying to schedule horticultural appearances.) In an effort to occupy her underemployed spouse, Leslie McCloud dispatches him to entertain a corrupt head of state from South America who is threatening to align himself with the Russians if, as threatened, his foreign aid is cut off. The First Gentleman and his honored guest chase cars around Washington, race speedboats down the Potomac, patronize a nightclub where they are beset by the charms of a stripper with the tabloid-ready name of Nana Peel, and end up on the front page of the newspaper.

Thad McCloud also discovers that being married to the president puts a decided damper on marital relations. Every time the gormless First Gentleman attempts to seize the moment, his preoccupied president—even if momentarily persuaded to succumb to his masculine command—is quickly seized by duty or summoned by a phone call from the secretary of state. The movie's most unflagging recurring joke is that becoming the leader of the free world leaves a woman no time, energy, and, especially, inclination for lovemaking. Madame President's new domestic policy is made clear immediately after her swearing-in ceremony, when she claims the presidential bedroom, furnished in dark woods and masculine prints, for herself and exiles her husband to the adjacent spousal chamber for a solitary night in its frilly four-poster. In *Kisses for My President*, Leslie McCloud gets her comeuppance, and the movie gets its resolution, when she discovers that—against all apparent odds—she has become pregnant. On her doctor's advice, and with a glowing if regretful smile, she resigns. Thad McCloud smugly notes that this outcome demonstrates the superiority of men: it took 40 million women to get Leslie McCloud into the White House; it took just one man to get her out of it.

Among the premises of *Kisses for My President* with which many among today's public might take issue is the suggestion that a pregnant woman is not fit to lead the nation. (Whether a woman who doesn't know how to avoid getting pregnant in the first place is fit to lead the nation is another question entirely.) But what the creators of *Kisses for My President* really failed to anticipate is that the first female presidential candidate would have less in common with the brisk, accomplished Leslie McCloud than she does with her suddenly dispossessed husband Thad.

In the light of Hillary Clinton's bid for the Democratic nomination, *Kisses for My President* looks not so much like an anticipation of a second Clinton administration as it does a retrospective of the first, in which a presidential spouse unaccustomed to being relegated to the domestic sphere is suddenly required to develop an overweening interest in decorating schemes and menu plans. (Even as First Lady of Arkansas Hillary had managed also to be one of the hundred most influential lawyers in the United States.) Mrs. Clinton's first few years as First Lady will be, of course, primarily remembered for the ill-fated efforts she undertook to reform the nation's health-care system. But much attention was also paid to her work with Kaki Hockersmith, the interior decorator from Arkansas, who redid the White House with swags and over-stuffed upholstery; and her hiring, in 1994, of chef Walter Scheib to overhaul the presidential diet. While Hillary Clinton was clearly a new kind of First Lady, political pragmatism demanded that she be the old kind of First Lady, too, working the second shift magnified to the power of a hundred.

Possibly Hillary really did care as much about curtains as she pretended to in those early months in the White House, which is just as well, since, if she is elected, it seems unlikely that Bill Clinton will be obliged to enact the role of bumbling nullity assumed by the writers of *Kisses for My President*. (Though menu planning, come to think of it, might be right up his alley.) Hillary's plausibility as a candidate is, in no insignificant measure, dependent upon the prospect of Bill reinventing the role of political spouse: being all that Hillary sought to be without attracting any of the censure.

The proposition at the comedic core of *Kisses for My President*—that conjugal relations within the first family go to hell when a woman occupies the Oval Office—has, in any case, been anticipated and outstripped by the comedies and tragedies that sex supplied during the first Clinton presidency. Kenneth Starr insisted that the nation be provided with a much closer than necessary glimpse within the intimate sphere of the Clintons' marriage; with luck the nation and the Clintons will be spared a reprise in a second Clinton presidency. Suffice it to say that the scenario depicted in *Kisses for My President*, in which Thad McCloud's efforts to tempt his wife into a postinaugural quickie on their first night in the East Wing are dispiritingly rebuffed, will be beside the point. Washington may not have slept here, but the Clintons already have.

Thirty-five years after *Kisses for My President*, it is a sign of progress that debates about Hillary Clinton's fitness for office have little to do with whether or not a woman president can also be a good wife and mother. That's less because those questions have been answered than it is because they need not be asked. Chelsea Clinton is too old to be making desperate phone calls to inappropriate boyfriends, begging to be spirited out of the White House; so the thorny question of whether a woman really can combine child-rearing, let alone childbearing, with being leader of the free world remains uncountenanced. As for whether a female president can be a good wife—well, that depends upon whom she is married to, and in this regard, William Jefferson Clinton is the ideal First Gentleman. Whatever else the electoral victory of Hillary Clinton would mean for the country—an inspiring leap forward for American women, a weirdly belated registering of the place of women in the professional marketplace, an alarming confirmation of the United States' regression to dynastic politics—it's impossible to imagine Hillary's election being anything other than her marriage's ultimate consummation.

9

MY GENERATION

THE THIRTY-YEAR QUEST FOR
SWEET AND STEELY

BY JANE KRAMER

What do I think of Hillary? The question is, really, *why* do I think of Hillary? Or, more accurately, why do I keep thinking about what I think of Hillary? I do not wake up thinking about Barack Obama or John Edwards or Joe Biden, and certainly I do not let Republicans trouble my sleep, since they are troublesome enough every waking day. But I am obsessed with Hillary Rodham Clinton. I am doing what men do, and what I swore never to do. I am turning a presidential candidate into a candidate for gender studies—and I don't know why. What I do know is that it has very little, if anything, to do with politics.

In 1970, I wrote a profile of a consciousness-raising group known in the women's movement as the first radical feminist cell in New York. Most of the women in it were a few years younger than I was then, and when they weren't arguing about why I wanted to have a baby—I was married *and* pregnant—they were arguing about power. They couldn't agree on whether power was good or bad, or even, really, about what it was. (They liked the idea of empowerment, but not the kind of empowerment that was carpet-bombing Cambodia.) They were angry about not having any power in a man's world, and, at the same time, they were certain that having it was not only corrupting but deeply shameful. They wanted access to power, but they didn't want to *want* power. They thought that ambition carried a Y chromosome, and with some reason, they blamed the mess in the world, and in their own lives, on that single strand. They liked sex—it was men that drove them crazy. They were marked by distress in ways that I liked to pretend I was not. They seemed to me then to have grown up without much sense of their own entitlement to whatever it was that men expected for themselves. They thought of themselves not so much as disenfranchised as inching their way to self-invention. They were all well-educated bourgeois women (the feminist vanguard was never proletarian), but the closest any of them got to political action was one woman—fresh from a violent marriage—who eventually entered law school and became a women's rights lawyer. They were on the cusp of Hillary Clinton's generation, but it was hard to imagine any of them delivering the kind of "women assuming power and responsibility" speech that she had given, a year earlier at her Wellesley graduation. They distrusted women who "got involved with power" or even with changing the rules of the game that people in power played. They had, in the end, a handmaid's view of women, and never mind if being a handmaid was precisely what fueled their anger. The few members of the group whom I have seen in all these intervening years are still angry. But why are they angry at Hillary?

There seems to be a common view among liberal American women—women like me—that it would indeed be nice to have a woman president, but there is an equally common view that electing a woman for that reason would be "essentializing," or, you could say, embarrassingly reductive. There is, concomitantly, a strong belief that a woman president might "do something" for women—as if, unlike the Y chromosome, the X chromosome was encoded with a special propensity for benevolence, some kind of super-justice genome—or, worse, an equally strong belief that she might not do anything for women, in which case she would either be "just like a man" or a selfish glass-ceiling striver (the "only woman in the room" syndrome) or treasonous to her sex.

Perhaps we should be thinking more about what the woman in question now has, or has not, accomplished with the power she already has. Surely, an argument can be made that Hillary Clinton squandered her truly unprecedented authority as First Lady on a health-care plan that had very little to do with making the country's doctors functional—not to mention making its patients healthier. And then there is her Senate performance of the past seven years. It does not take much political wit to see that all those votes and speeches have been pointedly at the service of her inevitable presidential campaign and, often, only incidentally at the service of the issues that, as the Democrats' reigning celebrity, she might have been expected to champion.

The truth, in the case of Hillary (condemned, like Arnold, to be called by her first name, like an odd, exasperating, or vaguely frivolous child), is that there is very little on record to suggest that she will do something for anyone but herself. She has been almost comically cautious and conciliatory as a senator. She has curried favor not only with the Republican old guard, or what's left of it, in the interest of reasonable compromise (which is, after all, what senators are supposed to seek), but also with the God-besotted men and women who may still make up the majority of the American electorate. She has tried to deflect the attention of the propagandists who gave us the Swift Boat campaign—to see it settle, perhaps, on Edwards or Obama. And who can blame her? That is what politicians do. It is certainly what her husband did. He caved on the appointments he wanted to make, and on the promises he wanted to keep. He was faithless in ways that had nothing to do with Gennifer Flowers or Monica Lewinsky. Even as a lame-duck president, he never said go to hell to a radical Christian right that even Hillary swore was the source of a conspiracy against him. The difference between them is that Bill Clinton wanted our love at least as much as he wanted our vote. Hillary Clinton wants us mainly at the polling station, and she lacks the rhetoric that might make us forget it—the needy, attentive, eloquent, irresistible Bill Clinton love pitch that we all miss so much.

Women, of course, are supposed to be better at love than men. That is the bottom line of essentializing. And never mind that it's hard to think of a female head of state who was much good at love, though there is a case to be made that Golda Meir ("the only man in the Cabinet," Ben-Gurion called her) loved her grandchildren. Hillary loves Chelsea and, it is clear, Bill, but the women who accuse her of manly ambition and coldness now—which is to say, of having no demonstrable love for *them*—also berate her for not leaving the man she has lived with for more than thirty years. In the event, the standards that women set for women seem to be different from the ones we set for men. Do we expect women to be "better"—morally, emotionally, socially on a higher level? Or are we simply resentful of a woman who says, in effect, okay, I want it all—and then steps over the rest of us to get it? Are we still, inexplicably, convinced that not wanting it all is what saves women from disgrace? Or are we simply falling into the classic liberal pattern of undermining our own best people? Why is the man who wants it all, at worst, unpleasant (and at best sexy), when the woman with the same hunger is a bitch?

The aims of feminism, at its origin, were civil and domestic justice—domestic justice being, at least in principle, enforceable once civil justice was established. And the assumption was that when that happened women on the march would shed their disabling diffidence and men on the retreat would shed their greedy, gloomy tyranny. We would embrace our inner males and they would embrace their inner females, and we would all live happily ever after, doing the dishes and running the world together. By now, those of us who have kept our heads (which is to say, not succumbed to today's new cult of genome literalism) should welcome ambitious women and domestic men. So why do we find Hillary unseemly in wanting political power or expect her to be discreet and seductive in pursuing it, as if the job description for a woman president read "courtesan with veto power"? (The real courtesan in the Clinton family is surely Bill, who, being always anxious to please, rarely exercised a veto.) It has been said ad nauseum that motherhood could be considered the most demanding form of leadership, calling for skills in salesmanship and negotiation and persuasion that are arguably beyond most of the backroom boys in Washington. The problem is that this is invariably said with condescension. It has also been said that politics today has devolved into a form of housekeeping, a kind of regulatory or diplomatic or administrative tidying-up of the house that money built. Power, the argument goes, is economic. It is the men of finance who make the world go round, and what we call politics can now be left to the good offices of a deferential and efficient wife. Men who believe this make a mistake. (Consider Angela Merkel, who took over as her party's housekeeper; swept the old boys into the dustbin, along with their deals and scandals; and, smiling sweetly, became chancellor. She is now quite comfortable with power, and demonstrably more cheerful having it.) Women believe it less. They know, or should know, that it is still the politicians who send their children away to invade countries the politicians don't like, and pack the courts with judges who want to punish them with abstinence. Hillary, to her credit, doesn't even pretend to believe it—which may be *her* mistake.

None of this answers the question of why I continue

to subject Hillary Rodham Clinton to the kind of scrutiny I would never think to apply to men. I look at the men running for president and ask myself if their politics are mine, or close enough to mine to be appealing. My interest in Barack Obama runs mainly to: Can you do the job? Are you brave enough for it? Do you have the vision for it? Can you take the heat? My question for Hillary Rodham Clinton is: Why do you want the job? What kind of woman does that make you? I take Hillary personally—too personally. I take her seriously. Perhaps my view of men is cynical. Or perhaps, in some ways, I know men better than I know women—or you could say, better than I know myself. I want Hillary to make a good impression. I do not mean her hair or the cut of her pantsuit; I want her to be a good, generous, and loving person *and* a steely, scary, effective person. I want her to change and not change with the job, while I expect, and maybe even accept, that a man will certainly change when a couple of hundred million people suddenly start calling him "the leader of the free world." I hope for the best, with men. I expect the best from women, and at the same time I know I will be disappointed. In matters of sweet and steely, I also disappoint myself. Maybe I have not evolved. The women's movement that I attached myself to, early on, is trying to figure out Hillary when it should be trying to win an election in 2008. I agree with Gloria Steinem. When asked if she was for Clinton or Obama, she replied, "I am for Clinton *and* Obama."

RE: HILLARY

Just because Hillary Clinton reminds me of every bad dealing I've ever had with a woman who's had power over me ~

every relative,

No, you may NOT have a cookie.

Because I said so.

every strict teacher, every unsympathetic employer, ...

I fail to see what is so amusing about Vasco de Gama.

You're four minutes late.

every annoying bureaucrat, every imperious saleslady, every bossyboots "Alpha" mom ~

If you do not have form 101-B, I cannot help you.

Don't touch that.

My Susie has a 98 average.

My Susie has never eaten at McDonald's.

I don't allow my Susie to watch the Simpsons.

10.

11

FATE IS A FEMINIST ISSUE

Has Hillary paid her dues from her own account?

by Judith Thurman

An old friend of mine in the film business likes to tell the story of his engagement, forty years ago. His fiancée, a ballerina, came from a conservative military family, so he did the proper thing and asked her father for her hand. Consent was given, albeit grudgingly, as the background and profession of this future son-in-law were, by the father's standards, unsuitable. My friend was too happy to care about, or perhaps to notice, the chilliness of his reception. "Now that we're almost relatives," he asked the old warrior, "what shall I call you?" "Call me General," he replied.

I think about the general whenever I watch Hillary Clinton work a room or give an interview. Her smile has a benevolent dazzle that seems quite genuine. It helps to dispel an impression common to her biographers and to many voters—that she is a prickly, opaque character wearing a mask of affability. More often, though, her voice and mien telegraph an unspoken rebuke or distaste for the public's presumption—a kind of lèse-majesté—in desiring to know her on more familiar terms, as if she were saying, "I'm not Hillary to you." However one addresses her, she deserves, or doesn't, to be elected president on her own merits, of which she has many, and not by virtue of her husband's. Yet Mrs. Clinton's strategic choice, complicity, or surrender in letting wifehood define her for the better and formative part of her adult life troubles and confuses many women—particularly, perhaps, professional women of her own milieu and generation, her natural base—who share her views on social justice and identify with her drive. We were the girls, and ours were the ideals (some naïve) that Hillary Rodham embodied in 1969 as Wellesley's first valedictorian, and as a law student at Yale in a class that was 90 percent male. We, and she, came of age during the civil rights movement and joined the vanguard of the feminist wave. For better or worse, we rejected our mothers' fatalistic acceptance or contented embrace of subordinate lives, and the well-meaning career advice of elders or guidance counselors that generally involved taking a secretarial course or a degree in elementary education, so that just in case we failed to catch a husband who was a "good provider" we could support ourselves respectably. And even though Mrs. Clinton set an important precedent by continuing to practice law in Arkansas, as the Rose Law Firm's first female partner, and boasted in her memoir *Living History* that "I represented a fundamental change in the way women functioned in our society," her trajectory—up to a crucial point—bears a striking resemblance to that of a woman who was her own mother's contemporary, Ruth Graham, the wife of the evangelist (another larger-than-life Bill) who died last June at the age of 87.

As a "twenty-one-year-old college graduate with a razor-sharp mind," Laura Sessions Stepp wrote in the *Washington Post* obituary, "Ruth McCue Bell made a decision that today might seem . . . somewhat peculiar. She planned to become a missionary in China, where she had spent her childhood. Then her good-looking preacher boyfriend asked her to marry him. 'Woman was created to be a wife and mother,' Billy Graham told her. Not I, she responded. But several agonizing weeks later she said yes," accepting the terms of a traditional partnership and a future of vicarious glory. The role of a pastor's wife "is one of the hardest jobs there is," Stepp continues. "Not only are you expected to obey and serve your husband, you're supposed to like doing so, and on the occasions you don't, keep quiet about it. Close friends are hard to come by because there is so much you're not supposed to discuss." As Mrs. Clinton discovered in 1992, the First Lady of a pious country founded by Puritans is still, at least symbolically, the pastor's wife, and she is revered as such only insofar as her own conduct is irreproachable, or her husband's isn't. Keeping quiet does not, obviously, come naturally to her, but for the sake of her marriage, and her fortune, she learned to do it.

In accepting Billy Graham's proposal, Ruth Bell sacrificed the autonomy of her ambitions. Mrs. Clinton, whose ambitions seem always to have been for the kind of worldly consequence to which her endowments of a steely temperament, methodical competence, and supreme intelligence suited her from youth, put hers on hold, investing her promise in the joint tenancy of an exceptional property—the presidency of the United States—to be recouped with interest at a future date. In that respect, her prescience as a young woman—not so much in understanding what she had to do but what she couldn't do, or couldn't be seen to do, on the path to power—suggests, ironically, the battle plan of another world-famous consort: Diana Spencer. As Tina Brown observes in her biography of the late princess, Diana (whose ambitions were steelier than one thought) cut herself from the herd of her competitors in the royal marriage tournament by forgoing many of the freedoms, revolts, opportunities, indiscretions, and experiments that girls born in the 1960s, as she was, claimed as a birthright. The job that she aspired to had certain archaic requirements, and the winning candidate had to be undefiled by the mistakes of a modern woman's life. One can't help but feel that in marrying Bill Clinton, Hillary Rodham was, with more calculation than perhaps is admirable, saving herself for her wedding night with the American people.

A queenly bow to social or religious convention in the service of one's country is not necessarily evidence of venality ("Paris vaut bien une messe" is the concession of a realist rather than a hypocrite), and Americans have not been averse to political dynasties, even if the last eight years have soured many voters from both parties on that notion. Some dynasts become great leaders. Many, if not most, leaders of both sexes have been supported by their spouses, their families, and a network of powerful connections. Bill Clinton was a self-made man, but it is hard to know how far his own gifts would have taken him with a different wife. And with a different husband—or none at all—Mrs. Clinton, given *her* gifts, might yet be serving in Congress, or on the bench, or as the president of a university or foundation. But she would almost surely not have been boosted into the high orbit she presently enjoys by the payload of fame and favor that she brought, at the age of 53, to her first campaign for elected office—as the senator from a state where she had no roots. It's not that Mrs. Clinton hasn't paid her dues, but rather that she hasn't paid most of them from her own account. Her official credit history in national politics starts in 2000.

That impressive if flawed and relatively brief history by no means accurately reflects the depth and breadth of Mrs. Clinton's professional training in framing policy, or in the day-to-day governance of a republic at the innermost circle of decision making as her husband's most intimate counselor and apprentice. But the problem in assessing the unique experience that she justly points to is its private, not to say clandestine, nature. After the derision that greeted Clinton's "two-for-the-price-of-one" campaign speech, and the debacle of Mrs. Clinton's national health-care initiative, her influence went undercover. She may have been the de facto co-president of the United States, operating, as no First Lady has before or since, from an office in the West Wing, and she is, in fact, running on the record of her contributions to one of the most successful and popular administrations in modern history—

but without accountability for her invisible role in its vindictive intrigues and myopic failures. Many New Yorkers, even those who may have voted for her, resented Mrs. Clinton's first Senate campaign, despite the posture of humility that she struck on her "listening tour," for what they considered her carpetbagging and opportunism, and if her creditable performance, particularly in consensus building, has since disarmed some of the criticism, the absence of notable legislative accomplishments or principled stands that might come back to haunt her with one or another group of swing voters has undercut her claims to moral leadership. Many Democrats currently deciding whom to support in the primaries are, despite their admiration for her political skills, put off by what they perceive as an overconfident sense of entitlement to the public trust that is disproportionate to her documented public service. Some of the animus against her "arrogance" is certainly generated by misogyny, and some, which harks back to the early, amateurish blunders of the first Clinton administration, is outdated, since Mrs. Clinton, who has a genius for adaptation, has learned from her errors and, despite her rigidity, has evolved. But the irony for feminists otherwise receptive to, if not thrilled by, the prospect of a woman president, is that only her husband may know how much of that trust is deserved.

HILLARY'S LIST

WHICH ICONIC LITERARY CHARACTER WAS HILLARY'S ROLE MODEL?

BY SUSAN CHEEVER

Personal reading lists are literary DNA, so it's not surprising that Hillary Clinton's list of favorite books is a double helix of balance and compromise. The seven books that Clinton enumerated for Oprah Winfrey on television a few years ago reflect her concerns as a woman and as a politician and her extreme concern about how to make those two things go together in a way that has never happened before in American politics. The bulk of the reading list is dominated by politically correct novels from feminist writers of three races and five nationalities. Alice Walker, Amy Tan, Ling Wu, Jean Auel, Barbara Kingsolver, Beryl Markham: this part of Hillary's list reads as if it was drawn up by a well-meaning contemporary Wellesley girl or a committee of eager-beaver staffers.

What *is* surprising is the first book on Hillary's list, a book that is something completely different. There's nothing politically correct about it. It's the only one on the whole list written before 1900. It's the only one that is one of the bestselling books of all time and famous worldwide, and it's the only one that could have meant anything to the young Hillary Rodham growing up in Park Ridge, Illinois, in a household that "resembled a kind of boot camp," according to Carl Bernstein's recent biography, *A Woman in Charge*. When one of his children left the cap off the toothpaste tube, Hugh Rodham, a former naval chief petty officer, threw it out the window and made the offender go outside to retrieve it, even in the snow. Under the fierce 1950s rule of this domestic martinet, the teenaged Hillary found refuge in reading long before she realized that even a list of books is a political statement.

The book at the top of Hillary's list is *Little Women*, written under duress in 1868 by the 36-year-old Louisa May Alcott in the second-floor bedroom of her parents' house in Concord, Massachusetts. Alcott wrote *Little Women* reluctantly, at the insistent urging of her own domestic martinet—a demanding nineteenth-century pater who ran a determinedly vegetarian, idealistic household boot camp of his own for his wife and four daughters. When Bronson Alcott was disappointed in one of his girls, she went without dinner; when he was *really* disappointed, *he* went without dinner and sat sternly at the head of the table with an empty plate while the rest of the family ate.

The literary odds were against Louisa May Alcott as she sat at the little desk her father had built for her at Orchard House and churned out a book about herself and her three sisters, a book that she found dull even as she was writing it. She was sick with mercury poisoning sustained from medication she had been given when she served as a Civil War nurse in the Union Hospital in Washington, D.C. She hadn't wanted to write a book for and about young women, but when her father—bribed by her publisher with a promise to publish a book of Bronson's if he was able to get Louisa to finish hers—ordered her to, she was unable to disobey. Yet although she was angry, unenthusiastic, and very ill, her experience and talent came together to create the spirited Jo March, an iconic character who redefined American femininity in what almost immediately came to be one of the most beloved and influential books ever written. The political odds are certainly against a former First Lady running for president from the U.S. Senate, yet experience and talent sometimes make the odds look silly.

All women have a choice of two paths in life. Will we depend on men and devote ourselves to pleasing them and building a domestic kingdom? Or will we try to make it on our own, developing our own careers and intellects and hoping that a man will find this intelligence and independence attractive? "When you come to a fork in the road, take it," Yogi Berra advised, and, like many of us, Clinton has chosen both paths. "We are, all of us, exploring a world that none of us understand and attempting to create within that uncertainty," Hillary told her own Wellesley graduating class at her 1969 commencement. "We're searching for more immediate, ecstatic and penetrating modes of living." Jo March was her role model. On Oprah's website, Clinton explains, "This book was one of the first literary explorations of how women balance the demands of their daily lives from raising families to pursuing outside goals."

Alcott, like Jo March, spent her life trying to reconcile her so-called masculine traits—ambition, intellectual curiosity, and a talent for leadership—with the nineteenth-century feminine ideal of wifely obedience, cooking skills, devotion to family, and passion for the art of hospitality. As a woman, Alcott thought she had more important tasks than baking apple pies with the red pippins her father picked from the orchard or entertaining neighbors at tea parties that avoided all discussion of politics. To express this impatience with the world of domestic graces in a way that was acceptable to her peers, she had to write fiction.

Hillary Clinton had a harder time. When she mused that, as the First Lady of Arkansas she "could have stayed home, baked cookies, and had teas," but that instead, as a lawyer and activist, she had worked to "assure that women can make the choices they should make," she might have been channeling the author of *Little Women*. The controversy caused by this unguarded statement is sad evidence that our culture hasn't changed much since the era of crinolines and spinsters. The right wing slammed Clinton as the yuppie wife from hell, and political cartoons showed Bill Clinton as a marionette with an angry-looking Hillary pulling the strings. Apparently all kinds of women were insulted by what came to be known as the cookies-and-tea remark. Hillary obediently receded; it took her a decade to build political power the old-fashioned way, for herself. Now, as the first wife and mother to be a serious candidate for president of the United States, she is also redefining what it means to be a woman.

Alcott describes Jo March as a plain girl who has a "decided mouth, a comical nose, and sharp gray eyes which appeared to see everything, and were by turns fierce, funny or thoughtful." It's a description that might fit the young Hillary Rodham. Like Jo, Hillary was introspective to the point of self-torture, bluntly curmudgeonly, and determined to devote her life to making a difference in the lives of others. "Since Xmas vacation, I've gone through three and a half metamorphoses and I am beginning to feel as though there is a smorgasbord of personalities spread before me," she wrote to her friend John Peavoy in April of 1967, when she was at Wellesley, in a letter which might have been written by the impatient, adventurous Jo. "So far I've used alienated, academic, involved pseudo-hippie, educational and social reformer."

I had lunch with Hillary Clinton once when she was still First Lady. She was thinking of running for the U.S. Senate and I was writing a parenting column. Long Island's Child Abuse Prevention Services—called CAPS—honored us together at a ladies lunch at the Marriott in Uniondale. As we leaned over our fruit salads, I asked her if there was anything I could do to help. There was something about her that drew me in. Is it politically incorrect to say that she was somehow extremely feminine? As we sat shoulder to shoulder and talked about children, I thought of Jacqueline Kennedy, and the way she drew you toward her with a soft, soft voice and then let her cloud of exquisite perfume, Joy, do the rest. Hillary was enchanting and impressive, and when she asked, "How can I reach you?" I felt that we had a connection.

"I know that we are a nation that cares for our children," she said that day from the lectern, "but all too often we forget about the children who need our care that aren't directly in our line of sight. There isn't any more important way that society is judged than how we care for our most vulnerable: our children, our elderly, and people who are otherwise left on the margin of society."

If Hillary becomes the first woman to be an American president, she will change our world; women have only had the right to vote since 1920. On the way there, though, Hillary Clinton will have been pummeled and dogged by her womanhood, by the difficulties of being a woman without being subservient, of being feminine without being domestic, of being sexually attractive without forsaking her intelligence. One of the most brilliant lawyers of her generation, Hillary chose to abandon Washington, where she had worked on the Watergate impeachment investigation, and go to Fayetteville, Arkansas, to be with her fiancé—a man who adored her but who was already having relationships on the side. As she creates policy statements on health care and the war in Iraq, the media reports on her cleavage and speculates that she wears pants to hide her legs. As she starts the exhausting trek from boondock to boondock in search of votes, the press goes on about whether she was a good mother and whether Bill has stopped cheating on her. Hillary, who once wore headbands and skirts, has had to suppress both her extreme masculine and feminine edges in a way that male candidates don't even know about. She has had to bake cookies after all and stand gracefully by her man, even as she has served in the men's club that is the United States Senate.

Hillary is always trying to get us to forget that she is a woman because she knows how distracting it can be. Yet it's the most important thing about her. She isn't just another suit. She's not one of those dead white males who still happens to be alive. Quietly, without tears or flirtatiousness, she is changing what it means to be a woman. "If she can win, women can do anything," a 12-year-old friend of mine said the other day as we picked up some cookies at the bakery. As the election that I hope will make her president approaches, Hillary Clinton seems to be relaxing into the feisty, smart, educated woman she was when she was younger, a woman who sounds a lot like Jo March when she grins and says, "I'm your girl."

CHEATING

DO MARRIAGE AND POLITICS MAKE LIARS
OUT OF ANYONE WHO TANGLES WITH THEM?

BY ARIEL LEVY

Cheating is easy. It's lying that's hard.
It is awfully difficult to look at the person you've loved most and longest and say something that just isn't so: "I was out walking the dog," "There was no cell phone reception," "I did not have sexual relations with that woman."

But the act itself? The slide from flirtation to connection to consummation . . . it's like riding a bicycle. (In heaven.) Your body gets old, gets married, becomes uninspiring, perhaps, to that other person who sees it night after night, year after year, but it never forgets what it wants. Try telling your body that you are serious about monogamy. It couldn't care less.

I can't think of a person I know and like who has not been cheated on or cheated or at least attempted to. And yet not only the hope but the expectation that good people will keep their hearts true and their pants zipped until death do them part remains intact. As Barack Obama has lethally noted, people have tired of prosecuting the questions of the 1960s.

But the question of how to reconcile the love of one's spouse with the relentlessness of desire has never been answered by any generation of freethinkers.

Maureen Dowd was perhaps the most prominent of the female critics who asserted that Hillary Clinton's failure to dump her husband after his philandering rendered her "unmasked as a counterfeit feminist." Dowd—among many—said Hillary had "let her man step all over her." But I'm not so sure. Perhaps Hillary's acceptance of her husband's pathological infidelity wasn't about being a doormat, as Dowd would have it, or about relentless ambition, as Jeff Gerth and Don Van Natta have suggested and John Podhoretz has declared. ("You and everyone you know probably despise Hillary Clinton," Podhoretz wrote in *Can She Be Stopped?* "You think she stayed in her marriage because she was hungry for unelected power, and that disgusts you.") Perhaps Hillary is guilty of being a broken-down realist—one who's learned time and again that everything in love and politics is a compromise.

After eight years of research, one of Carl Bernstein's conclusions about Hillary and Bill Clinton's thinking is that "the Journey—their term—was at once endlessly romantic and unapologetically ambitious, a trek across the political landscape in which they intended to inspire the expansion of the country's social consciousness, based on their own ideas and ideals, and those of their generation." I think a lot of people saw it that way. The Clintons were going to redeem authority, remake it into something that those of us who'd been made to feel degenerate and un-American (not least by a man who couldn't spell potato) could respect.

The first time I saw them I was stunned. Politicians, my parents had impressed upon me since childhood, were Them, not Us. Politicians were square, warmongering, establishment, full of it. Ronald Reagan and George Bush, the presidents I grew up with, had never given me cause to doubt this assessment. The Reagan and Bush administrations "couldn't recall" the details of selling missiles to enemies and funding right-wing Nicaraguan drug traffickers. They taxed the poor and wore tuxedos with the rich. They blamed the L.A. riots on Murphy Brown and the breakdown of family values and told us that Americans should be more like the Waltons. And those wives! Old ladies in pearls and pantyhose. They were everything my parents were not.

So when I went to a rally in Connecticut during my freshman year of college to see the Clintons, I couldn't believe it. I can still remember the feeling that went through me when Bill Clinton said, "If you're gay, if you're a single parent, if you're divorced," as my parents were by then, after two decades of cheating, "come join *our* family and share *our* values." It made me dizzy. The idea that everything could switch and the margins might become newly filled by Them and the halls of power presided over by Us—the radical, the righteous— well, I had never even contemplated the possibility.

When the Clintons first ran—and at least initially, they encouraged the perception that they offered a co-presidency with their "buy one, get one free" slogan—there was so much about them that was immediately familiar to 1960s people, and to the children of 1960s people, like me, who were voting for the first time in a presidential election. All that Fleetwood Mac and talk about social justice. Hillary declaring on *60 Minutes*, "I'm not sitting here, some little woman, standing by my man." She sounded like my mom. I didn't even think about the context (her defending his wayward prowlings), and it never occurred to me that her comments were impolitic. For those of us in families still clinging to the 1960s for a spine, it was just more of the stuff we'd been saying or hearing for years, only now it was coming out of the mouth of someone who mattered big time.

Also familiar, of course, were the flexibility and fallibility of the Clinton marriage, which came to our attention almost immediately. But so what: he'd cheated, and she knew about it. Shit happens.

CHEATING IS EASY. The real crime, the real humiliation, is not infidelity but indiscretion. You can't feel it when your spouse's body comes alive without you; you are mercifully unaware of what's happening in that hotel or that car or that Oval Office. Maybe your mate's ardor for you lessens as a consequence of the adultery, but maybe it's reinvigorated . . . or maybe the fizzling happened so long ago you don't even notice any change. There is no pain or humiliation in your beloved's body doing other things with other people. There is pain in finding out about it. There is humiliation if other people do.

Unless an affair is discovered, it is the tree falling in the forest that makes no sound because no one is there to hear.

It's lying that's hard. As fervently as Ken Starr labored to make the lurid particulars of Bill Clinton's sexual opportunism the focus of our disgust, the real horror was that he deceived us—the whole American family—and conscripted others to bolster his dishonesty. Donna Shalala and Madeleine Albright were like gullible aunts roped in to tell us, "No, really! Your dad's a good guy!" If only he'd settled with Paula Jones in the first place, we now know, confessed to yet another failure of restraint, so much more of his and Hillary's (and our) dream could have remained intact, unsullied. Cheating, unfortunately, almost always begets lying, and lying is contagious.

Lying is supposed to be for Them. Lying is for hypocrites, for people who don't know or don't want others to know who they really are. For politicians who don't understand what's actually important. To lie about cheating—rather than being smart and deeply circumspect, which is to say considerate, in the first place—is to reveal oneself as the worst of both worlds: parochial and deceitful.

After Bill Clinton was exposed as both a cheater and a liar, Hillary Clinton became the public's proxy—a walking Rorschach test for our feelings about infidelity. And she started having (even more) difficulty being honest. (Hillary, unfortunately, has borne out that dirty backstabber Dick Morris's assertion that she has "a complicated relationship" with the truth.) But while it's convenient (and conventional) to imagine that only a sucker gets cheated on, that only bad marriages encompass unfaithfulness, it's also delusional.

You can question Hillary Clinton's integrity based on her outlandish flag-burning amendment. (Please.) Or doubt her credibility because she's talked out of both sides of her mouth on the war in Iraq. As Carl Bernstein's own cuckolded ex-wife, Nora Ephron (who first became famous by chronicling her former husband's flagrant infidelity in her novel *Heartburn*), has written, Hillary "has taken the concept of triangulation and pushed it to a geometric level never achieved by anyone including her own husband." But to fault her for staying with a husband whom by all accounts she adored, who adored her back ("he genuinely worships that woman," in the words of former senator Robert Torricelli), with whom she believed she could change the world—and *has*—is folly.

Cheating is a constant, like love, that transcends political and generational categories. Remember what happened with Newt Gingrich? (Remember what happened with Zeus?) What is more worthwhile to wish for, in a spouse or a president, is accountability—something we're particularly hungry for in the United States after being denied it entirely throughout the current administration's tenure. For many liberals and feminists and flower children's children, this is the source of our real disappointment with Hillary Clinton, the first woman with a real shot at the presidency.

I miss the direct, brilliant, ball-busting Hillary who it was once possible to imagine existed. "While Bill talked about social change, I embodied it. I had my own opinions, interests and profession," Hillary wrote in *Living History*. "I was outspoken." In her current presidential campaign, Hillary's speeches have become as cloying and generic as that frigging Céline Dion song she chose by committee.

All of Hillary's competitors have the advantage of not having been publicly cheated on. Some have the advantage of not having had to vote on the war. Perhaps Barack Obama has the biggest advantage, simply because he hasn't been in politics long enough to cheat or lie, or, more significantly, get caught. It is hard not to wonder if, sooner or later, marriage and politics don't make liars out of anyone who tangles with them.

Love is rare, but desire doesn't care. It is as simple and impossible a puzzle as how to get elected and remain human.

THE VALIDATOR

Notes on a matriarchal upbringing

by Kathryn Harrison

In June of 1953, my mother's parents flew to Europe from the Los Angeles International Airport. I have a photograph of them, posed outside the Pan Am terminal. Fifty years ago, air travel held a glamour and momentousness that it has long since lost, and for the trip overseas my grandmother wore a navy blue suit, plum-colored high-heeled shoes, with a matching hat, handbag, and gloves. My grandfather was dressed in a summer-weight, dove gray suit with a pressed white handkerchief in his breast pocket. He wore a hat as well, a fedora with a pearly grosgrain band. Their appearance, soigné and almost leisurely in their calm—entirely lacking the frayed anxiety that has come to characterize air travelers, nervously anticipating endless lines at security checkpoints, unanticipated delays, cancellations, missed connections—belies the anecdote that goes with this snapshot, an account of the minutes that unfolded before the photograph was taken. The two of them were hurrying toward the terminal from their taxi, my grandmother in the vanguard as always, when my grandfather slipped and fell—in a puddle of oil, the story goes. By the time my grandmother noticed that he was no longer keeping up with her she was some yards ahead of where he was lying, on his back, on what he called the "tarmac."

"Harry!" she said when she looked back and saw him, and she stamped her foot. "What on earth can have possessed you to lie down at a time like this? Don't you understand? We're going to be late! You'll make us miss our flight!"

This image, of my grandfather lying on his back looking up at a cerulean (pre-smog) California sky while my grandmother stamps her foot and demands to know why he's lying on the ground when they're trying to catch a plane, is a favorite of mine. The story is embellished, no doubt, if not apocryphal—where are the suitcases? why didn't the cab drop them at the terminal door? how can his suit have remained perfectly and pristinely clean for the photograph?—but persists because of the truths it holds. I see my grandfather with his hands folded over his stomach, his jacket and trousers pressed, his hat still in place, wearing the same gentle, contemplative expression he almost always wore. My grandmother, a foot shorter and ten times fiercer than he, her dark brows drawn together, has a familiar expression of outrage on her face. The directions of their gazes are emblematic: his heavenward, toward the abstract and the ineffable; hers earthbound, fixed on her immediate environment and how she could change it. The two of them are revealed even in the colors they wear. In contrast to her vivid silhouette, his hair is white, his eyes a mild blue, his suit pale-hued. My grandmother kept her hair rinsed black, darker than her original dark brown, until, at 85, she let it lapse into a steely gray. But her lipstick remained red until she died, at 92, as did her cardigan sweaters.

My grandmother was—I suppose she would be called—a narcissist, although if there was ever a woman who defied conventional labels it was she. Certainly, she craved attention; she insisted upon it, at any cost. As a young woman she wasn't so much beautiful as radiant: vivacious and charming. She willed people to admire her, and mostly they did. The ones who didn't found her fascinating nonetheless. Older, she grew increasingly ruthless, and shameless, in her struggle to dominate those around her. When thwarted she screamed incoherently—not inarticulately, but without human language. She howled until her willingness to resort to what looked like madness, to play dirty, overcame her opponent.

I can't keep my grandmother out of mind when considering the possibility of a woman becoming president of the United States. It's reductive, and worse, irrational—what has my grandmother to do with Hillary Clinton?—for my response to a candidate's gender to proceed from my relationships with the people who raised me, but isn't everybody's sense of male and female rooted in family? Solidarity with my sex—and more, with feminism—makes Hillary Clinton's being a woman important to me. But it's an importance imposed by circumstances rather than conceived within my own heart. In 2008, and in all subsequent elections, I wish I could vote gender-blind. I don't need to elect a female president to shore up my sense of women's worth, already given me by my grandmother, who raised me for my young and irresponsible mother.

AMONG WHAT WAS peculiar about my family was the fact that it was securely matriarchal; all the power and energy, and the conflict, were invested in my grandmother. A man's labor and his wealth made this possible, as all of us—my mother, married and divorced by 19, my grandparents, and I—lived on the dwindling fortune left to my grandmother by her father. But it was easy to forget this, since he was absent, dead long before I was born, and the trust he established was in her name. When I was very young, allowed to observe but never to question matters relating to finances, I understood only that money, necessary for food and clothing and toys and every other thing, came to my grandmother, delivered quarterly by some mysterious and inexorable force. Somehow, from afar, she commanded coffers. As I grew older my understanding evolved, but the impression made so many years before remained: Nana had the money, and thus the power.

I don't know that she could have tolerated it any other way. Born in 1899 to a wealthy Jewish family who lived in Shanghai's international concession, my grandmother managed to endure matchmaking, but only as far as the altar, where she jilted the man who had been selected for her. It was only after her parents' death, when she commanded her inheritance, that she married, and at 42 she chose a man who was older, elegant, decent, and kind, uninterested in asserting his personality over hers. He was her superior, if judged on moral grounds; he was honorable and honest and gentle. She was impassioned; she burned bright; she loved drama. When it was lacking, she created it, warmongering with my mother.

My grandfather wanted peace. He wanted to get up in the morning, read the paper over his breakfast, and spend the morning gardening, which he always did in his suit pants and a shirt with buttons. He wanted to finish the paper over lunch, and then continue to garden until the five o'clock news. At six, he made dinner for the three of us and then retired to the den to read, a political biography, usually—the shelf behind his chair was filled with fat lives of Churchill, Eisenhower, Disraeli, and others whose names I can't recall. He enjoyed politics and kept abreast of minor developments as a baseball fan might study box scores. Although his formal schooling had ended at the age of 14, he read a great deal all his life and was by nature an intellectual. He liked to think an issue through, examine its varied aspects and the ways in which it might change depending on the circumstances. In his twenties he developed film and prints for a photographer, and I imagine his mental processes as similar to those of photography: in the dark, he waited patiently for nuances to reveal themselves.

My European grandparents, like most naturalized citizens, took their democratic responsibilities seriously. They paid their taxes; they remained aware of and grateful for freedom of speech and other civil liberties; they answered jury duty summonses promptly; they always voted. If away during an election, they cast absentee ballots. My grandfather, who took care to be informed on issues my grandmother was too lazy to study, told her for what and whom she should vote, and in most cases she followed his bidding, happy to be excused from the consideration of eventualities in a realm she didn't manage personally. As he was one of those very few people who could not imagine (and therefore never anticipated) dishonesty or corruption, he was often dismayed, disgusted, shocked to the core by what transpired in the nation's capital. I think he was as disturbed by Watergate as I was by 9/11; certainly he received it as the beginning of a new, dangerous, and profoundly worrisome era.

My grandmother paid little attention to Watergate. In that she possessed the sangfroid and ends-justify-the-means amorality of most successful tyrants, as well as the energy to champion a cause tirelessly until she won, I can imagine her a success in government, had her focus been less myopic. As it was, she didn't like to delegate, and the people she wanted to control, her family and friends, she wanted to control directly, one on one. I've kept her old address books, not so much as a record of her acquaintances but as evidence of her ability to judge without misgivings. Death and disenfranchisement were indicated by the same swift annotation, a hurried black slash from the lower left to the upper right of an entry, bisecting phone number, address, and name. No one who fell from her favor ever returned, not any more than those who had died. Photo albums reveal a similar urge: she decapitated individuals she came to dislike, excising only their heads from group portraits, leaving their bodies as evidence of their having been sentenced and dismissed.

I found her thrilling, of course. Not thrilling like my mother, whose power lay in her elusiveness, her ability to withdraw just at the moment I wanted her, leaving no more than a stylish outline and an ethereal whiff of Guerlain. But thrilling on her own terms. One aspect of my grandmother's personality that I value increasingly as time passes, and never more than during an election season, when politicians' pronouncements reflect the result of pollsters' advice and the careful crafting of speechwriters, was her authenticity, her complete lack of political correctness. Oh, she could be tactful, on those occasions when Machiavelli would have recommended tact, but she had no ability—or was it interest?—in appealing to whatever was deemed appropriate. The idea that public discourse would be guided by an imperative that no group, whether based in culture, race, gender, disability, or creed, be offended was one she would have found, well, offensive.

I'M UNCOMFORTABLE WITH the assumption—no, I reject the assumption—that Hillary's candidacy confirms women's worth. She is the first, and so it's inevitable that she be seen as representing her gender as well as her constituency. Because women like my grandmother have always influenced—directed!—the lives of individuals all over America, many of us regard a woman president as natural and inevitable. So why not wait until the right one comes along? Because while we are past needing her election to personally validate the competence of our sex, many of us feel a responsibility to elect Hillary, if possible, because of what her presidency would mean. Not to us, but to them: to anyone who might not agree that a woman could make as competent a president as a man, as could a black man or a Hispanic, a Zoroastrian, a homosexual. Our leader doesn't have to be a heterosexual white Christian male; it's time this truth is realized.

So, anticipating the Democratic primaries for the 2008 presidential election, I am paying especially close attention to Hillary, because she's a woman, and to Barack Obama, because he is black. I'd prefer to vote gender-blind, and race-blind, too. But it feels like a luxury, a thing I have to consider whether I can afford. Before anything else, I'll vote for whomever I believe can best address the disasters he or she will inherit, a person who might well be Hillary.

Whoever does become America's forty-fourth president will, like all people, combine aspects of the feminine and masculine. Coming of age in the 1970s, along with feminism, I was fortunate that the people who raised me presented atypical faces for their respective genders. My grandfather, secure in his male strength, was patient, gentle, and reserved. If he felt compromised by the fact that his wife's money paid the bills, he didn't betray it. An introvert, he preferred to stay home, with me, while my restless grandmother was out, lunching with friends, shopping, going to club meetings. He built a playpen on large rubber tires, with a canopy, and pulled me after him through his extensive garden. By the time I could speak I had watched countless hours of his quiet industry, and when I did my homework at the kitchen table it was he who looked over my shoulder, he who cautioned me to spend as much time and attention on the last math problem as I had on the first. He didn't speak unless he had something to say, and I don't remember his ever raising his voice, at me or at anyone.

My grandmother was charismatic, and she was feisty, argumentative. Drawn to light and drama and crowds—to any kind of stage—she was a performer with sophisticated and innate political skills. The very thing that made him uncomfortable, even unhappy, was the thing she sought: power over other people. She drove the car; he sat in the passenger seat. He made dinner; she made dessert and did the dishes. Born in 1890, he was raised by his widowed mother, who worked tirelessly to keep her children out of the poorhouse. Born in 1899, she was a daddy's girl who rejected the decorous, and decorative, example of her mother. I never had to be taught the truths of androgyny: that society comprised aggressive, controlling, "manly" women as well as gentle, retiring, "womanly" men. I grew up assuming that this was the way of the world. It was only in college, having left home and the small independent school I'd attended since I was 2, a school founded by a woman who looked and acted like a man and that rewarded achievement regardless of gender, that I discovered discrimination against women (and even then I only discovered it as a topic of study). The world has shown me sexism in practice, but I've persisted in my innocence—somehow it always seems a mistake, an exception, not a rule. Such was the influence of my upbringing, which has left me with just this much sense of responsibility: everything else being equal, I will vote for a woman over a man, because there are a lot of people out there, a lot of enfranchised citizens, who weren't as fortunate as I was.

THE WIFE, THE CANDIDATE, THE SENATOR, AND HER HUSBAND

Managing the transformation from consort to candidate

BY LETTY COTTIN POGREBIN

If 1992 was the Year of the Woman, 2008 is shaping up to be the Year of the Wife—which is not great news for Hillary Clinton. Although she's a candidate for the highest office in the land and no longer First Lady, Hillary still is a wife, and the standards by which wives are judged differ radically from the standards by which Americans typically judge their presidents. If she wins the Democratic nomination, every nook and nuance of her marriage will be subjected to intense media scrutiny. Whatever voters sense about the dynamics of her wifely role will reverberate subliminally, not just with women, who can't help but identify with the first major female presidential contender, but with men, who can't help but project themselves into the place of the First Gentleman. Regardless of how they feel about Bill or his politics, men are going to identify with him as a man and a husband, and this puts Hillary in a tough bind: As a candidate, she has to project strength and power. As a wife, however, she dare not do anything that might emasculate her husband.

One of Hillary's fundamental challenges in this campaign, therefore, is to reconcile the image of the ideal wife with the paradigm of "leader of the free world." Her transformation from consort to candidate requires a constant, delicate fine-tuning of both roles lest she violate one in pursuit of the other and be punished for it. It requires that she keep asking herself three questions: What must I do to make voters feel comfortable with my wifehood while allowing me to rise above it? How, given the deep cultural imprint of the traditional female support role, can I get people to accept me at center stage with my husband off to the side, polishing his Nancy Reagan gaze? And how do I position Bill to do me the most good and the least harm?

A CBS News poll conducted in June 2007 found that six in ten voters consider a presidential candidate's spouse a "very" or "somewhat important" factor. The poll did not differentiate between Bill Clinton and all the female spouses, a distinction that, I suspect, would have yielded more layered results. Nonetheless, the numbers are telling. A spouse is considered important by 65 percent of the women CBS polled and 50 percent of men. Republicans were more likely than Democrats and Independents (71, 54, and 49 percent, respectively) to say they would factor the presidential spouse into their vote. Among those 45 years and older, nearly seven of ten respondents said the spouse would affect their decision.

Unfortunately for Hillary, history suggests that a husband is a political liability for a female candidate. Or as one wise cynic put it, "If Bill Clinton got killed by a bus, Hillary would have one less thing to worry about." By midlife, men have had more opportunities to get embroiled in questionable business dealings or embarrassing encounters with the IRS. Geraldine Ferraro, Eleanor Holmes Norton, Jeanine Pirro, Dianne Feinstein, and Nancy Pelosi are among the female candidates who have had to explain, defend, or distance themselves from husbands accused of all sorts of financial and other improprieties. The most sterling woman can be tarnished by accusations against her husband, whether or not they prove true. In addition to the strain and humiliation of such an ordeal, she can be thrown off her game and lose time and focus from her campaign while mopping up after him.

Given the husband-as-liability premise, it's no wonder that while women make up only 17 percent of the U.S. Congress, they constitute more than twice that percentage of Congress members who are single, divorced, or widowed. Likewise, it's no accident that husbandless (some would say, husband-free) women like the secretary of state, Condoleezza Rice, the former secretary of state, Madeleine Albright, and the former secretary of health and human services, Donna Shalala have held senior government positions. Somehow, it's easier to run, win, get appointed, or serve in public office if you're maritally unencumbered.

As husbands go, Bill Clinton, even with his rock star

magnetism and unrivaled fundraising skills, happens to be a unique burden to his wife. First, no way is she not worrying about his zipper problem and the chances of a bimbo eruption between now and the election. Second, since his career has played out in the spotlight and since she's so closely identified with his record, she has to answer for his administration's flaws and failures. Third, it's tough enough to run "against Washington" when you're a sitting senator, but it's impossible to do so when you and your husband once ran the place. Fourth, a certain kind of voter believes that "the man of the house" will be the real power behind the throne, and when the house in question is the White House and the husband has already occupied the office his wife aspires to, such a voter can readily imagine him pulling her strings or bombarding her with coercive pillow talk. Fifth, there remains in the land at large an army of EOBs (Enemies of Bill) whose hatred of him, compounded by their loathing for her, fuels a fierce determination to prevent a Hillary Clinton administration (especially since she's likely to give him a plum post) and whose money funds a raft of opposition research and advertising.

Even more hazardous than a husband with enemies, whose habits and beliefs are known to millions, are the demands of wifehood itself. Traditionally, being a political wife has been a support role, a smile-adoringly-at-your-husband-while-he's-saying-the-same-thing-for-the-hundredth-time role. These familiar behaviors—refined into an art form by Nancy Reagan and Pat Nixon—are writ large in the rulebook for candidates' wives precisely because they mirror the marital dynamic that some Americans, regardless of their own spousal relations, still consider appropriate and "normal." Despite forty years of feminism and the commonplace phenomenon of women being both autonomous human beings and married, when a political wife steps out of the support role—Teresa Heinz Kerry comes to mind—pundits take notice, campaigns get nervous, voters react, and the candidate takes the hit.

If proof were needed of the impact of wifehood, one need only look to Ann Romney and Elizabeth Edwards, each widely judged to be a huge asset to her husband's campaign. Both have been credited with pristine personal histories, exemplary motherhood, unbesmirched loyalty to their husbands, general likability, and courage and candor about their illnesses (Ann's multiple sclerosis; Elizabeth's cancer). On the negative side, several candidates' wives have made various missteps. Though Hillary is a candidate herself and not "just a wife," she must be paying attention to these women's gaffes, both for their effect on her opponents' campaigns and for what they reveal about the minefields of wifehood, to wit:

- Judith Nathan Giuliani, the wife of Republican candidate Rudolph Giuliani, has been accused of lying (about her multiple marriages, nursing career, shopping habits); once working for a company that tested its surgical staples on dogs (or, as one radio host put it, "killing puppies"); calling her ex-husband, Bruce Nathan, "a kike"; as well as being spoiled, demanding, a drinker, a social climber, abrasive, jealous, and conniving.

It's a miracle that the family values party is willing to claim Giuliani regardless of who his third wife is. Giuliani stumbled when he said he'd want Judith to attend cabinet meetings and advise him on health policy since she was a nurse. Add his convoluted marital history to Judith's, plus the fact that they started dating while he was still married (and he called a press conference to announce he was leaving his wife) and you'd think the Republicans would want to throw him under a bus. Yet after the initial brouhahas, "Mr. Security" got a pass while his wife was excoriated in a profile in *Vanity Fair*.

- According to the *New York Times*, Jeri Kehn Thompson, the wife of Republican presidential candidate Fred Thompson, caused friction in her husband's campaign by "playing a central role" and being "involved in everything including his travel schedule, fund-raising focus and staffing decisions." The caricature of the interfering wife—we know the type, right?—eclipsed the possibility that, as a former Senate aide, political consultant, and spokeswoman for the Republican National Committee, Mrs. Thompson might actually have some expertise in these areas. Besides dubbing her a busybody, commentators seem distressed that she's "eye candy" (it's those gold sandals and formfitting gowns), twenty-four years younger than her husband (as if that's her fault), a "trophy wife" (i.e., young, beautiful, and accomplished in her own right), and a home wrecker (despite the fact that Fred was divorced for seventeen years before he married her).

- Michelle Obama appears to be the African American Hillary Clinton—a bright, achieving, articulate woman who sacrificed her career for her husband and family (for the moment, anyway). Michelle, however, has been criticized for dissing her husband, Democratic presidential hopeful Barack Obama, by ridiculing his big ears and "funny name" and publicly griping that he doesn't put his dirty socks in the laundry, can't make a bed as well as a 5-year-old, and forgets to put the butter back in the fridge. She's also been skewered for comments like, "He's just a man," or "Someday maybe he'll deserve all the attention," and for calling him "the brother" even though "it makes some white folks uneasy." Presumably intended to humanize the candidate, her remarks instead have marked her as "bitter," "pushy," "passive-aggressive," "man-bashing," "weird," "emasculating," and a "radical feminist." Standard operating procedure for a political wife is to build up her husband, but many observers, especially bloggers, believe that Michelle purposely has been cutting her guy down to size and making him look unpresidential. If Barack can't "rein in" his wife, they say, how is he going to control Osama bin Laden or Kim Jong Il?

- Besides being disliked for her fortune (she's chair of Anheuser-Busch, which was founded by her father), Cindy McCain, wife of Republican candidate John McCain, is still shadowed by her 1994 admission that she was addicted to the prescription drugs Percocet and Vicodin. This revelation was made messier by news of a federal investigation into her theft of painkillers from the nonprofit medical relief organization she ran. What's most troubling about the story to me is not that she hid her habit from her husband (that's what addicts do), but that when she was ready to come clean, she had a lawyer call him, as if she was afraid of what he'd do if she told him herself. Juxtaposed against all those stories about John McCain's monumental temper, this sounds worrisome.

As for Bill Clinton's job in defining Hillary's role as a wife, one would think that he'd already proved his mojo by running the country and therefore would be free to be a supportive husband without fear of losing macho points. But masculinity is a fragile reed and often, without feminine buttressing, it wilts. One slip on Hillary's part—say, an open mike that catches her scolding him or ordering him around—and she'd be cast as a castrating bitch in a heartbeat. To carry off wifehood with grace, Hillary will have to convincingly convey that she adores and admires Bill but is not ruled by him. He'll have to be a beneficent presence in her campaign but not overbearing or spotlight-stealing; solicitous but not wimpy; agreeable but not henpecked. From Hillary's end, the key to this balancing act is to convey the couple's mutual respect—and, unlike many political strategies, this one has the virtue of being rooted in truth. By now, the country surely recognizes that both Clintons are whip-smart and ultracompetent, so it makes sense that they would respect each other. I maintain that we won't give a fig if they've been plotting since Yale Law School to take turns being president as long as we can believe that their partnership is genuine (which, it happens, I do). Many married people identify with or aspire to an equal partnership themselves. Women appreciate men who respect them, and men feel comfortable supporting a strong female candidate if she triggers associations with analogous women whom they already respect—a wife, if they're lucky, or a smart grown-up daughter, an able colleague, a revered female teacher, or, increasingly, their doctors, lawyers, or clergywomen. As long as Hillary doesn't come across as a ball-buster or challenge her husband's dignity, she's the perfect candidate to inspire those favorable associations. If you're a Christian conservative, male or female, who believes that women should submit to and be ruled by their husbands, you won't buy into this perspective—but then you're probably a long shot for the Hillary camp anyway. However, if you're a modern young couple trying to establish an egalitarian relationship, if you're a single woman hoping for a partner who will appreciate her strengths and not make her fake girly-girl weakness, if you're an older couple struggling to work out the kinks in your marriage and sick of rigid sex roles, then a clear model of respectful but nondeferential wifehood will resonate for you.

Of course, this strategy presupposes that Hillary is able to strike a pitch-perfect note on the femininity scale, which, as any woman can tell you, is not easy. Psychologists have found that traits considered feminine—acting subservient, submissive, delicate, passive, dependent, helpless, or needy—correlate closely with traits associated with wives. So here's another bind. Hillary has to be believably *wife-ish* while studiously avoiding any behavior that might be construed as *wifely*. Delicate and submissive are not part of the job description for commander in chief.

Other female heads of state offer scant guidance on successful wifehood. Margaret Thatcher never had to brave the rocky shoals because, from all indications, her husband, Dennis, was content to remain firmly in the background. He was a self-assured gentleman who refused interviews, made brief, bland speeches when asked, and referred to his wife as "The Boss." At no time during Thatcher's eleven years in office was either her persona or Dennis's masculinity compromised by unflattering gaffes or embarrassing revelations. Golda Meir, the prime minister of Israel, was deeply domestic—famously offering homemade coffee cake to visiting dignitaries—but though her bio said "married," her wifehood was long atrophied. It was common knowledge that Golda and her husband Morris were, for all intents and purposes, estranged. Still, divorce would have been politically untenable in a country whose culture, both Jewish and Arab, puts great stock in marriage and family. The simple solution was to keep Morris out of sight. Chancellor Angela Merkel of Germany has had to make excuses for her publicity-shy husband Joachim Sauer, who keeps himself out of public view. Sauer, who skipped his wife's inauguration, has been called "dour," a "killjoy," an "inept socializer," and detrimental to her public career.

If a husband is a liability for a candidate and wifehood is a normalizer for a woman, the ideal status, obviously, is for a woman to be a widow. Gloria Steinem often quotes Margaret Mead's line to the effect that in a patriarchal society, only widows are honored in authority, and indeed, an amazing number of female presidents and prime ministers have assumed power after their husbands died or were assassinated in office. Widows have ruled Bangladesh, Guyana, Sri Lanka, the Philippines, Nicaragua, and Panama, among other countries. The widow enters the fray with the sympathy of her constituency, a sense of entitlement, the assumption that she will advance her husband's legacy, and the credential of having been a wife without the complication of having to accommodate the demands of a living man.

Other leaders who are husbandless include Kim Campbell, who was divorced at the time she was briefly prime minister of Canada, in 1993; Ellen Johnson-Sirleaf, the president of Liberia, also divorced; and Michelle Bachelet, the president of Chile, who is legally separated. Among unmarried prime ministers have been Maria de Lurdes Pintasilgo of Portugal, Eugenia Charles of Dominica, and Hanna Suchocka of Poland. American women, too, have been appointed or elected to public office as a result of their husbands' deaths. To name a few—Wyoming's Nellie Tayloe Ross, the first woman governor; Hattie Wyatt Caraway of Arkansas, the first female senator (appointed and then elected twice more in her own right); two noteworthy women from Maine, Margaret Chase Smith and Olympia Snowe; Muriel Humphrey of Minnesota, widow of Hubert; Maureen Brown Neuberger of Oregon, widow of Richard.; Huey P. Long's widow, Rose McConnell Long of Louisiana; California Congresswoman Mary Bono, widow of Sonny Bono who died in a skiing accident; and New York congresswoman Carolyn McCarthy, who ran for Congress after a madman killed her husband. Most recently, in the Fifth Congressional District of Massachusetts, Niki Tsongas, widow of senator and presidential candidate Paul Tsongas, won the special election to Congress, becoming the first woman to represent Massachusetts in the U.S. House of Representatives in more than thirty-five years. So the pattern holds: widowhood greases the track, wifehood comes with potholes.

Nonetheless, I'm counting on Hillary Clinton to keep Bill out of the bus lanes and get herself elected anyway.

ELECT SISTER FRIGIDAIRE

Who cares if she's a phony?

by Katie Roiphe

I have yet to meet a woman who likes Hillary Clinton. They may agree with her politics, may think that she would be an effective leader, may support her candidacy for president, but they don't like her. Most express this unfortunate state of affairs with a resigned and mildly regretful air. "Like," of course, is a slippery and complicated word; it shields us from the responsibilities and revelations of our preferences. It allows us to hide from our own feelings: it is as mysterious and ineffable and outside of ourselves as physical attraction or love. What can we do? Would that it were otherwise! But we just don't *like* her. We like her husband, but we don't like her.

We don't know her, of course, though we know people like her. To be fair, our impression of her comes from a conglomeration of newspaper clippings and television appearances and photographs in our heads, a hologram, a fantasy, a mystical conjunction, a quirky coalescence of a million tiny experiences of our own with the news. When Bill Clinton was running for president in the early 1990s, I was still in school, and everywhere there were fashionable badges and buttons: Hillary for President. In those days, the outlandish, sly slogan had a certain allure: the improbable, humorous point, the flashy rhetorical gesture. And yet now that there is a distinct possibility that Hillary could in fact *be* president there is a marked lack of enthusiasm surrounding the prospect.

Years ago the *New York Post* ran a column entitled "Just What Is It About a Phony Like Hillary Clinton That Makes My Skin Crawl?" And the word "phony" is the key to a certain strain of animosity against Hillary. It is no longer a revelation that politicians are phonies, that the perfection and refinement and deployment of phoniness is in fact politics at its best. Vast swaths of the country are entirely conversant in the language of spin, completely at home with the constant attempt to describe the mechanism of phoniness, the creation of political character that is everywhere in the news. In fact, integral to the entertainment of the political scene is watching the aides and consultants spin. And so: why should it matter, why should it make our skin crawl, that Hillary Clinton in particular is a phony? Or rather, what is it about her specific brand of phoniness that irks us?

The distrust that many express toward Hillary nearly always returns to the vexed and unresolvable question of her relation to Bill Clinton. The central manifestation of her phoniness appears to be her marriage, which many persist in viewing as an "arrangement," a word which began cropping up as early as the Gennifer Flowers scandal. The implication is that Hillary is so interested, so pathologically invested in politics that somewhere along the line she made a "deal" that she would tolerate her husband's infidelities. A Republican strategist once referred to the Clintons' marriage as a "merger" and this sense of the deal, of the business transaction, of the chilly conglomeration of powers, lingers in the public imagination. When Hillary offended a country singer and her fans by saying on *60 Minutes*, after the Gennifer Flowers incident, "I'm not sitting here, some little woman standing by my man like Tammy Wynette," she in fact offended a much broader group by the implication that she was not tolerating her husband's affairs just because she felt the same sort of pathetic, self-effacing love other more ordinary women feel; she was tolerating them because of their *shared political goals.*

In fact, for a brief time, it seemed that her husband's womanizing could be a perverse gift to Hillary's public image. As the Lewinsky scandal broke, Hillary enjoyed a surge in popularity. She appeared on the cover of *Vogue*. She was dignified, yet hurt, a stance we seem to enjoy in our first ladies. But when she started talking about the "vast right-wing conspiracy" one morning on the *Today* show, she dipped in likability again. She was back on message. The personal was political. This seemed a jarring and unforgivable swerve away from the story that was supposed to be about intimate and recognizable things like betrayal and pain. Was politics all this woman could think about?

In the wake of the Lewinsky scandal, there was much speculation about whether Hillary was somehow complicit in her husband's affairs. The summer after the impeachment hearings brought emergent rumors that Hillary faked being mad at the president, that she wouldn't hold his hand on her way to a vacation in Martha's Vineyard as part of a deliberately staged effort to appear angry. She was pretending to be hurt because she knew it would play well with the American public, that it would humanize her. While this scenario seems wildly improbable, the fantasy itself is revealing: the idea that she may not have suffered from Clinton's infidelities was much more disturbing than the idea that she had. A widely circulated anecdote from the sensational book, *State of a Union: Inside the Complex Marriage of Bill and Hillary Clinton*, by Jerry Oppenheimer, fed into similar suspicions. It described a letter she allegedly wrote to Clinton before they were married: "I know all of your little girls are around there. If that is what this is, you will outgrow this. Remember what we've talked about. Remember the goals we've set for ourselves. You keep trying to stray from the plan we've put together." The lingering sense that she might have signed on for the life that she in fact lived was, for some reason, unforgivable. Did she know that he was going to cheat on her? Did she choose to marry him anyway? The possibility that she was in a marriage whose narrative was not centrally about love informed the image of her as cold, inhuman, a virago, a *phony*.

Indeed the question looms over every book about Hillary: how motivated by power and ambition is she? Precisely how detached is she from what we consider the normal human emotions? Hillary herself comments wryly on this when, in *Living History*, she writes, "some people were eager to see me in the flesh and decide for themselves whether or not I was a normal human being." A certain hardness, an independence, an ambition in all her endeavors, has always struck observers. Her high school yearbook predicted that Hillary Rodham would become a nun called "Sister Frigidaire." And this sense of her as mannish, cold, clung to her. One of the latest Hillary books, *Her Way: The Hopes and Ambitions of Hillary Rodham Clinton*, by Jeff Gerth and Don Van Natta Jr., includes a dark, paranoid account of the "plan" in which each of the Clintons occupy the White House for eight years. And the image of the "deal" continued to shadow the discussion long after Bill Clinton left the White House. When Hillary ran for Senate it seemed to many as if this too might be part of the "deal." This prospect is often cited as part of her fakery, her sham marriage, even though it seems that this sort of deal, on more minor levels, and in more subtle ways, takes place in marriage all the time.

As a thought experiment, let us say that Hillary was in fact all the things that she is accused of being, that all of the most sinister accusations against her involving "plans" and "pacts" were true. Let us say that the idea is that she is using her marriage as a vehicle for power, that she was from the beginning attracted to Bill Clinton because she knew he would bring her closer to the center of power. Let's say she was one of those people for whom love, erotic attachment, and all of its attendant pain was secondary to her desire to run the world. Let us say that all of what she herself calls "the brittle caricatures" of her are true. Why should unnatural ambition be so alarming in a presidential candidate? Why should the single-minded pursuit of power at the cost of all personal relations be so unlikable? Why shouldn't we want Sister Frigidaire for president?

In fact, Hillary's drive, her ambition, her hard work, her deft manipulation of power, her refusal to be vulnerable, her unwillingness to allow love to get in the way of career goals, at least in her mature years, could be seen, if anything, as a sign of strength. She is in many ways the feminist dream incarnate, the opportunity made flesh, the words we whisper to little girls: "You can be president. You can do anything you want." Surely if one had said to a group of women waving picket signs in the 1970s, one day there will be a presidential candidate as ruthless, as cold, as willing to sacrifice relationships for power as any man, they would have been heartened. And yet, even our admiration for her undeniable achievements has a chilly aspect, an abstract, pro forma quality. If Clinton is in many ways the embodiment of certain feminist ideals, then it may be that many of us don't like feminism in its purest form.

It is interesting to note that in spite of predictions to the contrary, Hillary has a much more comfortable relation to younger, blue-collar women, a much more effortless popularity. It is, paradoxically, the women most like her, the demographic most similar in their education and achievements, that have the most difficulty with her. This is curious. It makes one wonder whether there is an element of competitiveness to the dislike, a question beneath the surface: why her and not me? Strong, accomplished women who one would think would respect her, would identify with her, may in fact resent her. Could it be that we like the idea of strong women, but we don't actually like strong women? There is an intolerance on the part of powerful women toward other powerful women, a cattiness, a nastiness, that is not a part of any feminist conversation I have ever heard. It is so much easier, so much cooler, so much more appealing to have a Hillary for President button when Hillary is not, in fact, running for president.

There is also the matter of Hillary's forced relation to femininity. Her transformation from a woman who cut her own hair and wore work shirts and jeans and no makeup to a coiffed blonde in pink cashmere and pearls has been much noted and commented on. Hillary writes in *Living History* about the arrival of stylists in her life during Clinton's first campaign: "I was like a kid in a candy store, trying out every style I could. Long hair, short hair, bangs, flips, braids and buns. This was a new universe and it turned out to be fun." But of course one doesn't get a sense of fun from that particular passage. Here again is the phoniness. Her pleasure and mastery of traditional femininity is not effortless; rather, one feels the labor, the artifice. At one point, *Time* magazine accused her of "allowing handlers to substitute the heart of Martha Stewart for her own." And it seemed that way because her relation to all things female felt unnatural, contrived. What is interesting is that this groping for a kind of workable femininity, a palatable, mainstream feminine image, necessary as it was, bothered us. Hillary was unable to project the effortless image of a strong, yet feminine woman that the next generation, at least, has come to expect. It is meant to be easy, to be seamless, the transition from tough, serious workaholic to lady in kitten heels. It is meant to look natural, and the sheer awkwardness, the effort that Clinton projected, the contradictions that she so conspicuously, so crudely embodied were perhaps a little close to home. She is trying too hard, and the spectacle of all this trying is uncomfortable, embarrassing. One could feel in a palpable way the smart woman's impersonation of the pretty woman, the career woman's impersonation of the stay-at-home mom; one could feel a lack of grace. This is perhaps the quandary of our feminism, how hard we are trying to be both, how often it feels faked. Hillary's "phoniness" may be so irritating, so unforgivable, to so many smart, driven, women in part because it is our own.

We may vote for Hillary come Election Day, but will we ever like her?

FROM THE 1965 *EYRIE* YEARBOOK
(Maine South High School, Park Ridge, Illinois)

BY PATRICIA MARX

HILLARY ("TAKE OFF YOUR GLASSES, LET DOWN YOUR HEADBAND!") RODHAM

First, second, and third highest GPA average (all four years). Committee to invade New Trier High School (chairman, years 3, 4). Subcommittee to Stay Out of New Trier High School (chairman; years 3, 4). Park Ridge Thespian Society: *Macbeth* (roles: Lady Macbeth, Macbeth, year 2); *Hair* (coiffure understudy, all roles, year 3); *Anything Goes (But Don't Quote Me)* (legal affairs, year 1). "Limo for School Bus Scandal" (denied involvement, year 3). Suspended for cheating on home ec. exam (hid chocolate-chip cookie recipe in cheeks). Broke into Vinnie

F.'s house and did his homework (allegation never proved). Little known fact: tattoo on ankle says: "There's a Methodist to My Madness." Waffle Club, all four years). Varsity Whitewater Rafting (captain, years 3, 4). First prize, honors essay: "Hey, How Come Girls Can't Wear Slacks?" Task Force on Ways to Reform Nurse's Office (spearheaded Universal Exemption from Gym Proposition 23a). Invoked cloture on student council vote to enforce proper spelling of cloture. Brought Robert's Rules of Order on (all three) dates. Aspiration: two children—Chelsea and Midtown; destruction of the Berlin Wall. Hobbies: conducting focus groups, arranging flowers (specialty: *genniferus*). Pet peeves: bad punctuation, martial law. "It takes a village to lift her thousand-page term paper." "On the one hand. On the other hand." "Stay as cunning as you are!" Never forget the great times auditing the Key Club Bake Sale!

SENIOR STATS

Favorite saying: "What does the Vast Right-Wing Conspiracy Club know anyway?"

Favorite animal: chameleon.

Favorite color: opaque.

Favorite food: anything frozen.

Favorite cosmetic: my trusty Cover-Up stick.

Voted "Student Least Likely to Be Married Someday to a Charismatic, Noninhaling, Philanderer Who Feels Other Peoples' Pain"

Voted "Most Popular Unpopular Girl"

THE DOUBLE BIND

The damned-if-you-do, damned-if-you-don't paradox facing women leaders

by Deborah Tannen

In researching a book about women and men at work, I spent several months taping conversations, observing interactions, and interviewing people at a corporation. When the book, *Talking from 9 to 5*, was published, I gave a lecture reporting my findings at the corporation, after which I asked audience members to write reactions and comments. One of the responses included a remark that surprised me and made me feel vaguely criticized. A man who worked in an office where I had spent a great deal of time commented that he noticed that I'd had a makeover. (I hadn't.) I thought of this when I read the title of one of the many demonographies that have been written about Hillary Clinton: *The Extreme Makeover of Hillary (Rodham) Clinton*, by Bay Buchanan. The concept of a "makeover" captures a lot about the way many people respond to women in positions of power, and to Hillary Clinton.

Women in authority are subject to a double bind, a damned-if-you-do, damned-if-you-don't paradox. Society's expectations about how a woman should behave and how a person in authority should behave are at odds. If a woman speaks and acts in ways that are expected of a woman, she will be liked but may be underestimated. If she acts in ways that are expected of a person in authority, she may be respected but will probably be viewed as too aggressive. The characteristics that we associate with authority are also characteristics that we associate with men. This doesn't mean that every man possesses those traits, but if he aspires to, his path is clear: anything he does to fit society's image of a good leader will also bring him closer to society's image of a good man. For a woman the two goals conflict. To the extent that she fulfills the expectations associated with being a good leader, she violates those associated with being a good woman. If she meets expectations associated with being a good woman, she veers away from the characteristics we expect of a leader. That is the essence of a double bind: anything you do to serve one goal violates the other. And that is the vise that has Hillary Clinton—and potential voters—in its grip.

I observed this dilemma in my research on women and men at work, and I wrote about it in relation to Hillary during Bill Clinton's first presidential campaign. At the time I dubbed it "the Hillary factor." Here's how it worked then: Hillary started out paying little attention to her hair, holding it off her face with a headband and leaving its natural color unchanged. For this she was ridiculed. (She wasn't conforming to what society expects of a good woman: paying a lot of attention to her appearance.) So she did what her critics seemed to demand: she got her hair styled and highlighted. Then she was ridiculed for trying several different hairstyles as she sought one that worked. (She wasn't conforming to what we expect of a public figure, which is steadiness and consistency.) Then her new hairstyles became fodder for interpretations of her character: she was too changeable, too concerned with appearances. In a word (though the word wasn't used at the time), she was pressed to have—then blamed for having had—a makeover.

The concept of a "makeover" is deeply associated with women. For one thing, it is nearly always women who undergo makeovers, which entail changing aspects of appearance that are more malleable—and more closely watched—in women than in men: hair, clothes, and makeup. In my case, all I did before I gave my lecture was put on makeup, as I always do when I'm speaking in public. But when I made my daily visits to the corporation's offices, I didn't wear makeup—a practice pretty standard for women in academia but unusual for women in the corporate world, so it probably was noticed. You can't say about a man, "He didn't wear makeup." The statement has meaning only in contrast to expectations. So no man need make a decision about wearing makeup—a decision no woman can avoid, even as she knows that whatever choice she makes will incur the approval of some and the disapproval of others.

I also dressed more formally when I gave my lecture. A man could certainly improve his appearance by dressing more stylishly and getting a better haircut, but the change is likely to be less dramatic because the range of styles available to men in these arenas is narrower than the range available to women. Most men have hair and clothing styles that are neutral, so their choices rarely lead to interpretations of their character. A man can choose a style that will, like a ponytail or a comb-over, but the point is, he doesn't have to. There is no such thing as a neutral style of hair or clothing for women. The range from which a woman must choose is so vast that any choice she makes will become the basis for judgment about the kind of person she is. And there will always be people who think her choices could be improved. (That's why mothers and daughters so often critique each other's hair and clothes—each wants to help the other look her best, though such suggestions are usually perceived as criticism.) How often have you looked at a woman and thought, "She'd look better if her hair were longer/shorter/curlier/straighter/with bangs/without bangs/with different bangs/a different color," and so on?

The most striking thing about the concept of a makeover—and it is key to understanding how people react to Hillary Clinton—is the notion that there is something slightly manipulative about having one, as if a woman is unfairly wielding the weapon of appearance to achieve her goals. That's why it hurt my feelings a bit to learn that someone thought I had had one. That's the Hillary factor. Hillary is subject to these kinds of contradictory reactions to almost everything she does. When she first assumed the role of governor's wife in Arkansas, she was disliked for being too independent: she kept her maiden name as well as her career. When she began behaving more as a political wife is expected to, taking her husband's name and attending more "wifely" functions, poll results indicated that people felt she didn't have an identity of her own.

Hillary has also been vilified for not acceding to demands that she apologize. For example, many people have angrily urged her to apologize for having voted to give President Bush authority to invade Iraq (although she emphasized, when casting that vote, that the authority was to be used only after all diplomatic options were exhausted). But, in other instances, she has been vilified *for* apologizing. After the failure of her health-care initiative, Hillary invited a group of women journalists to the White House to talk about how she had been portrayed in the press. Although she believed that the gathering was off the record, one journalist taped the proceedings and later reported that Hillary said, "I regret very much that the efforts on health care were badly misunderstood, taken out of context, and used politically against the administration. I take responsibility for that, and I'm very sorry for that." This triggered a barrage of attacks. One newspaper quoted a political scientist saying, "To apologize for substantive things you've done raises the white flag. There's a school of thought in politics that you never say you're sorry. The best defense is a good offense." A Republican woman in the Florida state cabinet was quoted as saying, "I've seen women who over-apologize, but I don't do that. I believe you negotiate through strength."

The assumption that apologizing is a sign of weakness

applies to public figures of both sexes. But apologies typically work differently for women and men. Research shows that women tend to say "I'm sorry" far more often than men do. But a woman's "I'm sorry" is often not an apology; it can be an expression of regret, shorthand for "I'm sorry that happened." Clearly that's the spirit in which Hillary made her remarks about health care. Ironically, the criticism she faced for "apologizing" reflected just the sort of misunderstanding and being "taken out of context" that her remark lamented. It's the double bind again: Anyone in the public eye is likely to resist apologizing so as not to appear weak. But a man who does so is fulfilling society's expectations for how men should behave. (Indeed, many men do fulfill these expectations, judging by how many couples' arguments revolve around the woman's frustration that the man won't apologize, and the man's frustration that the woman demands an explicit apology.) In contrast, when a woman in public life resists apologizing, although she is doing what is expected of a politician, she is flouting the expectations associated with women.

Moreover, a man who apologizes, as John Edwards did for his vote on the Iraq War, is starting from the position of strength that characterizes our images of authority and of men, so he can afford to sacrifice a coin from that stash of symbolic capital. But a woman who apologizes reinforces the assumption that women are weak. Any woman in public life must overcome this assumption, because weakness is at the core of our understanding of femininity. Many, if not all, of the ways that men have traditionally been expected to behave toward women—such as being chivalrous and protective—are based on that notion. In fact, it's the source of a double bind confronted by men. Let's say a man tries to be polite by holding doors open for women. If he doesn't, a woman might protest: "Didn't your mother teach you any manners?" But if he does, a woman might also protest: "I am capable of opening a door, you know." It's a double bind because a man who doesn't observe traditional politeness rituals risks offending by appearing rude, but if he does observe them, he risks offending by implying that women are weak.

All human relations pivot on two intertwined dimensions: on one hand, closeness/distance, and on the other, hierarchy. We ask of every encounter: Does this bring us closer or push us apart? And also: Does this put me in a one-up or a one-down position? Researchers in my field refer to these dimensions as solidarity and power. You can see them at work in forms of address. If you call others by their first names, you're exercising solidarity, bringing them closer; using their titles and their last names creates distance. But forms of address also operate on the hierarchy dimension, especially if they're asymmetrical. Calling people by their first names can indicate their lack of power, as with children and workers in service roles. Addressing someone as Mr., Ms., or Dr. can indicate either formality or the fact that they are above you on the social ladder and, hence, more powerful. This constitutes another double bind, and it brings us back to Hillary.

Why, we might ask, do we refer to Hillary as Hillary? Women are far more often referred to by their first names than are men in similar roles. This is partly because people tend to feel more comfortable with women and find them less intimidating. For a political candidate, that's a good thing. But being referred to by first name is also the result, and simultaneously the cause, of women commanding less respect. During the Democratic primary campaign debates, Hillary shared the stage with Kucinich, Edwards, Biden, Richardson, Dodd, and Obama—not Dennis, John, Joe, Bill, Christopher, and Barack. Of course, one obvious reason that Hillary is Hillary to us is that she shares her last name with the other famous Clinton (a choice, recall, that was pressed upon her). Another is that her name is unusual and therefore more recognizable than, say, Susan or Mary. But the name Barack is even more unusual. Without thinking it through, people are likely to view someone with whom they are on a first-name basis as less authoritative.

When we speak, we unthinkingly choose words that come to mind, but often the words we choose affect our thinking. The linguist Robin Lakoff pointed out years ago that the words commonly used in relation to women are often different from those used for men, and they typically create an impression of weakness. Lakoff noted, for example, that men may "pass out" but only women "faint." Many of the words used to describe Hillary are far more often used in connection with women. Take the list that Michael Tomasky quoted, in the *New York Review of Books*, from one of the many anti-Hillary screeds: "cold, bossy, stern, and controlling." Try out these words on male candidates, and ask how much sense they make. Pairing adjectives and candidates at random, let's see: McCain is cold. Giuliani is bossy. Bloomberg is stern. Thompson is controlling. Like the phrase "didn't wear make-up," adjectives implicitly contrast what is described with what is expected. Men are less often called "cold" because women are expected to be warmer, so the level of aloofness required for that adjective to apply to a man is far greater than the level required to apply it to a woman. The word "stern" suggests particularly female stereotypes, such as the schoolmarm. And "controlling," though it can be used to describe women *or* men, is, I think, more often used to describe women—and is more pejorative when applied to them.

The word "bossy" is particularly gender-specific. Think of children at play. Girls often reject other girls for being "bossy," while boys rarely apply that term to other boys. Indeed, a boy who tells other boys what to do is the leader. But girls value the appearance of sameness and tend to reject a girl who puts herself above others by telling them what to do. I encountered this ethic in my workplace research.

When I asked supervisors what makes a good manager, the answer I heard most often from women was, "I treat the people who work for me like equals." (The answer I heard most often from men was, "I hire good people, then get out of their way.") The expectation that a woman should not tell others what to do, or act as if she knows more (even if she does) or is higher up the ladder (even if she is) puts women in any position of authority—or in politics—in a double bind. If she does her job, she's bossy. If she doesn't tell others what to do, she's not doing her job.

There's another trap that frequently ensnares Hillary: the post hoc attribution of intention. Although we all know how often things happen to us that we didn't plan, we nonetheless tend to assume that what happens to others was plotted, sometimes nefariously, in advance. Hillary is widely perceived as ambitious: she must have plotted her ascent to the Senate and to the presidency. (Never mind that it's on the public record that the idea of her running for the Senate was first suggested by New York congressman Charles Rangel.) Here again we face a double bind. Ambition is expected of men who hold high office, but it violates our expectations of a good woman.

Isn't any individual who seeks public office ambitious— just like anyone who seeks a promotion or applies for a coveted job? But in a woman ambition is assumed to be a failing rather than a prerequisite. This explains another pattern I noticed during my research: women often believed that they should not make it obvious that they wanted a promotion—that is, appear ambitious—but instead should just do a great job and assume that it will be noticed and rewarded. (This conviction turned out to be a liability; promotions typically went to those who asked for them, or who already behaved as if they had the higher position.)

Yet another paradox makes life tough for women who stand out from the crowd. Studying junior high school girls, the sociologist Donna Eder made an unexpected observation: popular girls are widely disliked. At first this seems counterintuitive, but when you think about it, it makes sense. A lot of girls want to be friends with a popular girl in order to raise their own social status. But no one can have that many friends, so the popular girl has to rebuff the advances of most other girls. Hence, she's perceived by many as a stuck-up snob. I wonder if this isn't somehow at work with Hillary. Grown-up girls want her to be *their* best friend, to promote exactly the policies and beliefs that they espouse. And they can't forgive her when she doesn't. She must have another best friend.

This is one of many ways that Hillary seems to be held to a higher standard—a gender-based pattern that I noticed while writing a book about mothers and grown daughters. In talking to women, I quickly observed that many people expect more of their mothers than their fathers, and more of their daughters than their sons. Fathers, sons, and brothers may be busy at work, but mothers, daughters, and sisters should be there when you need them. For example, a woman commented that she barely notices when her sons don't call for weeks at a time, but if her daughters don't call for a week, she becomes concerned.

A related pattern that emerged in my research on families is that women are easier targets. The sociologist Samuel Vuchinich recorded sixty-four family-dinner conversations and tallied who started conflicts, and with whom. Children, he found, initiated more conflicts with mothers than with fathers. This finding reminded me of an observation made by a friend who sails competitively: early in the race, he said, he looks for ships skippered by women, because they'll usually give way more easily. This means that, whether or not the women skippers indeed behave as he expects, they are the ones he is targeting.

A small incident that received a brief flurry of attention in the summer of 2007 epitomizes a fundamental double bind that underlies all the others. Writing in the *Washington Post*, Robin Givhan reported that as Senator Clinton spoke about education on the Senate floor, the neckline of her blazer revealed cleavage—a revelation that Givhan described as a "small acknowledgement of sexuality and femininity." The very notion of cleavage encapsulates the questions that women face all the time: Is she sexy enough (to be a good woman)? Is she *too* sexy (to be a good woman—in a different sense of the word "good")? Since few people agree where the line is drawn between too sexy and not sexy enough, women are always at risk of—and sure to be seen by some as—violating one standard or the other.

Givhan noted that as a senator, Hillary "had found a desexualized uniform: a black pantsuit." Now, every male senator wears a "desexualized uniform." But we wouldn't say that about them just as we wouldn't say they don't wear makeup, because men in professional contexts aren't expected to dress in ways that call attention to their sexuality. But sexuality is inherent in our concept of femininity, so failing to display the former violates the latter.

For a stunning illustration of this double bind, consider the story of Joan of Arc. To lead the armies of France against British invaders, Joan dressed in military garb—that is, as a man. Then, dressing as a man became one of the charges that led to her being burned at the stake. In the trial scene of George Bernard Shaw's *Saint Joan*, the charge is levied like this: "she wears men's clothes, which is indecent, unnatural, and abominable." In her own defense, Joan explains, "I was a soldier living among soldiers. I am a prisoner guarded by soldiers. If I were to dress as a woman they would think of me as a woman; and then what would become of me? If I dress as a soldier they think of me as a soldier, and I can live with them as I do at home with my brothers." In the same way, when Hillary dressed in "a desexualized pantsuit," she simply dressed as a senator.

When Joan says that the soldiers "would think of me as a woman," she means as a sexual object. She didn't have to say that because sexuality is entailed by the word "woman." That's why dressing as a woman invites thoughts of sex—thoughts that turned into an article in the *Washington Post.* Any reference to a woman's sexuality highlights her vulnerability and hence compromises her authority. But, like Joan, a woman today who dresses in a way that does not display sexuality (that is, like men in similar positions) will be burned (only figuratively, thank goodness) for failing to be feminine.

Not everything about Hillary and the way she is perceived reflects the fact that she is a woman. But an awful lot does. Though we rarely think we are reacting to others in particular ways because of their gender, it's pretty hard, maybe impossible, to see anyone except through that prism. It is important for businesses to take these double binds into account when evaluating women and men for hiring or promotion, to ensure that they accurately assess employees' abilities. It's even more important when it comes to accurately assessing the abilities of candidates for public office. If we want to see Hillary for who she is and what she can do, we need to look more closely—not at her, but at how we look at her. Perhaps it's our own understanding of the double binds she faces that requires a makeover.

FIRM HILLARY

HOW THE CULTURE OF CORPORATE LAW
SHAPED HILLARY

BY SUSAN LEHMAN

S tiff, bloodless, sexless, excessively concerned with the *appearance* of impropriety, lawyers are not much fun at parties. This is particularly true of corporate lawyers. They are professionally risk-averse; relentlessly, *unreasonably* reasonable; people who look perpetually ready to pull another all-nighter in the library stacks. And: they are trained to be exhaustively, expensively argumentative. The culture of corporate law is one in which rigorous attention to detail is richly rewarded, and in which no detail is more rigorously attended to than victory.

The female corporate lawyer is a special breed. Her clothes nearly say it all: *I am buttoned up. My attempts at personal expression are awkward and sort of silly.* Remember when female lawyers wore man-tailored shirts with bouquets of fabric that bloomed just below the chin? Today's corporate costume makes a more straightforward statement: *I am here to do business. I am in it to win. I will eat your entrails for breakfast (and bill you for it), but I will not stain my Thomas Pink blouse as I do so.*

As Michael Tomasky, the author of *Hillary's Turn*, put it, books about Hillary Clinton, like mosquitoes on the tidal basin, arrive seasonally and in profusion. So do theories about Clinton, about who she is and how she came to be that way. An inarguable fact—and lawyers love inarguable facts!—is that Hillary Clinton spent the longest stretch of her professional life working in a corporate law firm. From 1977 to 1992, she worked as a lawyer in the firm of Rose, Nash, Williamson, Carroll, Clay, and Giroir (renamed Rose Law Firm in 1980) in Little Rock. She devotes a single sentence to these years on her campaign website: "She continued her legal career as a partner in a law firm." (And this, in a section called "Mother and Advocate.")

It's hard to imagine that the hours, days, nights, weeks, and years that she worked with a small group of white men on behalf of big businesses, banks, and brokerage firms had no effect on the singular phenomenon that is Hillary Clinton. In many ways, she is a product of corporate legal culture. I mention this not because it says anything about whether or not she would be a good president, or even whether she would be a good party guest, but because it may have something to do with the trouble we—or at least I—am having warming up to the fierce-minded, breathtakingly competent woman who keeps telling us she's "in it to win." And I am not the only one having a tough time liking, to say nothing of enthusiastically supporting, a woman who is obviously smarter, better informed, more focused, and more committed than most of the rest of us.

The Rose Law Firm is the ultimate establishment firm. It was established in 1820, before Arkansas joined the Union, and also before the state bar banned women from joining—a ban that lasted until 1912. Rose describes itself, in all official and promotional materials, as the oldest firm west of the Mississippi. (It is not clear why this is a matter of distinction, though Rose certainly was, as Clinton herself said, in *Living History*, "venerable.") It has always served the state's most powerful corporate and banking interests, including Arkansas's three biggest employers, Wal-Mart, Tyson Foods, and Stevens, Inc.

The Rose Law Firm was not the sort of place that the girls at Hillary Clinton's 1969 Wellesley graduation would have expected her to end up. In the commencement speech that she delivered (and which got her written up in *Life* magazine), Clinton bashed "competitive, corporate culture"—"not for us," she said—and described her Wellesley cohort's collective longing for "immediate, ecstatic and penetrating modes of living."

How much immediacy, ecstasy, or penetration Hillary found at the Rose firm is open to debate. Law firm life isn't exactly a laid-back kind of project in which one is in danger of losing oneself in the bliss of the moment. But for Clinton, the first woman ever to practice at Rose, the pressure and intensity of corporate existence had to have been of an entirely different order. She was, from the start, under constant scrutiny. Several of the firm's nine partners opposed the idea of hiring a woman lawyer in the first place and raised questions about possible conflicts. "How will we introduce her to clients?" an associate asked Web Hubbell and Vince Foster, the partners who had recruited Hillary and became close Clinton associates. And: "What if she gets pregnant?"

Secretaries gossiped about her frumpy clothes and her issues with her weight. Even peripheral members of the Rose community kept watch over her and felt free to intervene when she crossed what they considered to be lines of proper conduct. The first time Hillary had lunch with her colleagues Foster and Hubbell, a troublemaker called the men's wives to say they had been seen in a restaurant with "a woman." In a culture governed by principles of *stare decisis* ("the past rules") Clinton was, simply by being there, a source of attention and anxiety.

It is hard to imagine that Hillary didn't know that she was being closely monitored and viciously mocked. Put in a similar situation, most people would crawl under a rock. But that's not what Hillary did. Instead, she went to work. She continued to lunch regularly with Hubbell and Foster at the Lafayette Hotel, and was duly promoted to partner. She found her way onto the board of directors at TCBY, Arkansas Children's Hospital Legal Services, Lafarge, and Wal-Mart, where her position as the company's only female director made her become the object of even greater scrutiny.

If you look at a photograph of the 1990 Wal-Mart board, you see, front and center, in the style of a "Which of These Things Is Not Like the Others?" puzzle, Hillary, sitting among fifteen white men in dark suits. Though she is the only one in the crowd smiling (a genuine, open-mouthed smile), it is hard to imagine that she could actually have been that relaxed. Hillary billed thousands of hours during the years she spent at the Rose firm, and she also did substantive pro bono work, co-founded Arkansas's Advocates for Children and Families, and worked for the Legal Services Corporation, where she went on to become the first woman chairman of the board. Given all this, it seems possible that she has not relaxed for decades, possibly not since the words "ecstasy" and "immediacy" left her lips on the lawn at Wellesley in 1969.

Had an entire office and outlying community not been keeping close watch on the Rose Law Firm's first lady lawyer, Hillary still would have had good reason to keep careful watch over herself. The corporate law firm creates in its employees a state of hyper self-consciousness about their time and its value. The peculiar invention—some would say defining principle of corporate legal life—the billable hour, insures that this is so.

In most firms, associates and partners must account for every hour of their time, often in six-minute intervals. Day-dream, dilly-dally at the water cooler, linger over lunch: that's time that can't be accounted for. Lost time. Time without value. The time sheet is the firm's equivalent of the eyes of T. J. Eckleburg, which peer over ash heaps in *The Great Gatsby*, a kind of all-seeing conscience that creates a rigid, exacting self-regard.

The billable hour and the habits of mind it generates promote not just a souped-up sense of self-worth, but also what many lawyers describe as a perpetual state of anxiety. A friend who practiced in a big Chicago firm likes to say that it is a culture of disappointed father figures: *Could you do more? Are there precedents that could be unearthed? Additional arguments that could be foreseen—and countered? Hours that could be billed?*

As a welcome gift to Clinton, Foster and Hubbell gave Hillary a copy of Dickens's *Hard Times.* A dark denunciation of materialist culture, the book was an odd choice for the new Rose associate. (*Bleak House*, Dickens's tale of an interminable lawsuit, seems a more obvious choice.) It's tempting to guess that, by giving Hillary Dickens's critique of capitalism, Hubbell and Foster were winking at the kind of internal ethical conflict that corporate practice might stir up for their new associate, whose only other law experience consisted of a summer at Treuhaft, Walker and Bernstein, a radical constitutional and civil rights firm in Oakland, California.

The ability to argue all sides of an issue is a hallmark of the lawyerly mind. Hillary's ability to assert moral residency on different ideological sides of an issue showed itself soon after she joined the Rose firm. ACORN, a community organizing group that works on behalf of the poor, had helped pass a local ballot initiative that gave low-income residents a break on utility bills and increased rates for businesses. Wanting to put a quick stop to this handout, the business community called upon its lawyers at the Rose Law Firm and asked them to defeat the ordinance in court.

Hillary soon found herself battling ACORN's founder—and her close friend—Wade Rathke, in court. Deftly marshaling constitutional theory, she convinced the judge that the ordinance constituted an unlawful taking of property. The law was nullified. Wade Rathke never spoke to Hillary again. For Hillary, though, the matter seemed entirely impersonal: she has maintained strong connections with ACORN and works with them today, on minimum wage and election reform issues. She fought with and defeated a friend in court. That's business. That's what lawyers do.

One thing Clinton did not do at the Rose firm was spend much time litigating. For the most part, she stayed out of the courtroom and tried only a few jury trials. In the first of these, she represented a canning company that had been sued by a man who found rat parts, specifically rear-end rat parts, in a can of pork and beans. He couldn't think about the rat or the can without spitting, he said, which made it difficult to kiss his fiancé. So he went to court seeking damages from the canning company. Clearly something had gone wrong, but Hillary argued that the plaintiff hadn't really been damaged, and besides, the rodent parts had been sterilized in the canning process and might, in other parts of the world, thus be considered edible. She won. The plaintiff was awarded nominal damages—and she endured inevitable jokes about the "rat's ass case"—but Hillary, in a very rare public statement of inadequacy, said that arguing in front of the jury made her nervous.

I'm not sure when the Little Rock Courthouse was built or what its acoustics are like. But I know that courthouses often have high, sometimes vaulted ceilings, which create acoustical problems for women, whose voices tend not to project in the booming way that those of male lawyers can. To be heard, women lawyers have to raise their voices in ways that make them sound strident and shrill, an accusation that is frequently leveled at powerful women in careers in the law and beyond it. In any case, despite a successful track record with juries, Hillary Clinton stayed out of the courtroom, and even today seems deliberately to keep her voice low and well-modulated.

Neither Hillary Clinton nor the average corporate law partner is likely to make anyone's blood jump or their heart sing. When you are in trouble, however—real trouble—it may be that the person you want to see isn't the guy who wows you with his wit and charisma but someone who has really done her homework, pored over all the boring details, and then gone back over them again, just for fun. It's pretty clear that the country is in real trouble. Bridges are falling down; the stock market is all over the place; and let's not even bring up Iraq or Sudan. This might or might not be the right time to look past Hillary Clinton's cool, corporate, bill-by-the-hour sensibility; her lawyerly inclination to avoid risk and run everything past the pollsters, to smile and keep a stiff upper lip because appearance and propriety matter more than most things—and certainly more than impropriety.

MEDUSA FOR PRESIDENT

WHAT HILLARY'S MALE BIOGRAPHERS REVEAL
ABOUT THEMSELVES

BY LAURA KIPNIS

*"The sight of Medusa's head makes the spectator stiff with
terror, turns him to stone."*
—SIGMUND FREUD, "MEDUSA'S HEAD"

Medusa, too, had hairstyle issues. She's often represented as having snakes for hair; anyone who beheld her was turned to stone. Then there was that unsightly decapitation, at the hands of Perseus. In a brief essay on the myth, Freud comes up with some complicated theories about how the sight of a horrifying woman can be simultaneously transformed into a symbol of reassurance for a man—at least for men with anxieties about their manhood. (And what man doesn't have them, at least in the Freudian universe?) On the one hand, "to decapitate" means "to castrate," but in Medusa's case, according to Freud, the sight of the snakes actually *allays* the horror of castration, since they symbolically replace what's feared to be missing. Indeed, Medusa's decapitated head was often emblazoned on Greek shields as an emblem, since the stiffening fear to which Freud refers is a consolation—when going to war, when faced with terrifying women—reminding the bearer that he's still the one with the penis.

Consider a modern example, one from the realm of electoral politics. After all, the specter of loss certainly looms at the moment, at least for men. A woman is running for president and the small matter of who runs the world and how power is divided between the sexes is up for grabs in a far starker way than it's ever been. How could this *not* create a degree of anxiety at some unspoken and perhaps not entirely conscious level?

So what gets spoken of instead? Well, *hair* for one thing. Among Hillary haters—particularly the men, and particularly men on the right—her changing hairstyles have come in for quite a bit of commentary. But really, the very sight of Hillary Clinton, from hair to ankles, is a problem. She's monstrous, gorgonlike; not feminine enough, or conversely, deploying feminine wiles to further her nefarious ambitions. She might become president, and something about that possibility turns certain men stiff with outrage. To be sure, there are female Hillary haters too, and let me say at the outset that I'm not particularly a Hillary fan myself (I don't like her politics). But I'm going to venture one of those dangerous generalizations about gender: I think most women understand the impulse to try out different hairdos. Her male critics, however, can get positively obsessive on the subject, sending insults and projections flying left and right like shrapnel. Let's take a look at what they reveal in the process—not about Hillary, but about themselves.

In Edward Klein's bestselling *The Truth About Hillary: What She Knew, When She Knew It, and How Far She'll Go to Become President* (2005), the attention paid to Clinton's physical appearance is so microscopic that you fully expect to turn the page and find an index of her moles, accompanied by a close reading of what they indicate about her moral insufficiencies. Which are, according to Klein, legion. This is one despicable lady: a backstabber with a hard-left agenda—inauthentic, narcissistic, plus she has tacky taste in interior design. She experimented endlessly with her hair because "it was the one thing she *could* control about her looks," which were problematic to begin with. Though given her "penchant for secrecy and concealment," not to mention all the outright lying, you get the sense that her hair experimentations were merely the outward sign of a general preference for deception. This seems to be Klein's gist. At the same time, Hillary had "always thought of herself as an ugly duckling," we learn, and particularly hated her body, which caused her to neglect her personal appearance as a young woman, and to go around dressed like a hippie in shapeless clothes and hair that looked like it hadn't been washed for a month. In fact, she felt "so hopelessly unattractive that she did not bother to shave her legs and underarms, and deliberately dressed badly so she would not have to compete with more attractive women in a contest she could not possibly win." (Here I feel compelled to note, having seen a few photos of Klein, that this is a man who can't have felt entirely secure about his competitive mettle on this score either, hence, one imagines, all the snideness.) It wasn't until Bill ran for president that Hillary started to get her look together, having her hair teased into "a mighty blonde helmet, à la Maggie Thatcher" (elsewhere we hear that Madonna is her "sister in blonde ambition"), though as a busy junior senator, she once again stopped paying attention to personal grooming and started looking "like a zombie: baggy-eyed, gray-skinned, zoned out from lack of sleep," her hair hanging limp around her temples.

In addition to his discerning eye for hair volume, Klein is a connoisseur of the female form. Though Hillary is "a small-boned woman from the waist up, she was squat and lumpy from the waist down, with wide hips, calves, and ankles." (Klein refers to Hillary throughout the book as "the Big Girl," savoring the double meaning.) There's a medical explanation for her lumpiness, however, since Klein quotes an "anonymous medical authority," who speculates that Hillary may have contracted an obstetric infection after giving birth to Chelsea that may have resulted in chronic lymphedema, a condition that causes "gross swelling in the legs and feet," which Hillary covered up with wide-legged pants. Forgetting that this diagnosis is speculative (and, as far as I can tell, nowhere else confirmed), Klein goes on to say that lymphedema contributed to Hillary's preexisting self-image issues, and all the running around she does as a senator further inflames it, causing her ankles and legs to swell still more, making it painful for her to walk. Still, soldiering on despite the self-image problems and bloated feet, Senator Clinton soon adopted a "new high-pro-file strategy," discarding her somber black pantsuits in favor of brighter, more telegenic colors, and acquiring a chic new hair-do. Additionally, the skin on her face was now pulled tight, as though she'd had Botox. "As indeed she had," Klein chortles, trotting out another anonymous medical source (or perhaps the same one), who confirms that Hillary has been "Botoxed to the hilt." It must be said that Klein isn't entirely unflatter-ing in his depiction of Hillary, allowing that she's "the kind of homely woman whose features seemed to improve with age." Though he quickly adds that this "remarkable physical trans-formation" was a perk of her tenancy in the White House, since "as First Lady, Hillary received the kind of personal care that was available only to A-list stars in Hollywood."

Madame Hillary: The Dark Road to the White House (2004), by R. Emmett Tyrrell, is so titled because Hillary re-minds Tyrrell of Madame Mao, the "white-boned demon" who was never more dangerous than when wearing a seduc-tive guise. (Elsewhere he concedes that his American madame was also once a Goldwater girl who worked as an intern for the House Republican Conference.) Like Klein, Tyrrell grants that Hillary's looks have improved (he's even courtly enough to say that she's evolved into a "handsome woman"), at least since those youthful days of granny glasses with "Coke-bottle lenses" and baggy clothes. Like Klein, Tyrrell is another as-pirant for the Vidal Sassoon endowed chair on the Clinton-hating right. We hear, variously, that when Hillary first arrived in Arkansas, her hair was greasy and swept back under a cheap headband; that once in the White House her hair continued to be "problematic," causing her to "run through scores of appalling coifs"; and lastly, that "her search for the perfect hairstyle has finally been resolved into a neatly elegant businesswoman's coiffure" and she "seems to have turned her hair into a major strength." Additionally, her "thick eyebrows, which once would have collected coal dust in a Welsh mining village, are now well plucked and shaped into pleasant curves." Hairdo, eyebrows—thankfully we're not privy to data on the condition of her bikini line.

Along with the new do and makeover, Hillary has lost weight, observes Tyrrell, making her round face more angular—yet also somehow softer. (Facial softness is apparently a positive attribute in the Tyrrell lexicon.) Her eyes are bright and observant, Tyrrell notes, and "when she recognizes someone she likes, her face breaks into an expression of controlled glee, with wide-eyed, open-mouthed delight." Elsewhere, however, he charges that when she laughs, her eyes don't move. This is because she's "a surface phony," so much of one that she's even changed her eye color—from hazel to baby blue—through the use of tinted contact lenses. Or so a "Clinton insider" told Matt Drudge: "She even tried turquoise contact lenses once, but it was not a great look for her." Tyrrell's final verdict on Hillary: though her disposition is dark, sour, conspiratorial, and corrupt, still "one must acknowledge and even admire her strength, the indefatigable will at her core." Nevertheless, the election of Hillary Clinton to the White House would mean "a left-wing rampage," and Tyrrell doesn't mean the Democratic left; he appears to think she got her political education at the feet of Pol Pot. In a rather strange passage toward the end of the book, we learn that Hillary's ultimate dream is to be commandant of a "national Cambodian re-education camp for anyone caught wearing an Adam Smith necktie or scarf." Or perhaps it's also an extermination camp, since he adds: "Welcome to Camp Hillary. Please remove your glasses and deposit them on the heap. (Was that a flash of gold I saw in your teeth?)" I think Tyrrell means this to be witty. He concludes by telling readers he's "taking the high road, since hatred is an acid on the soul."

David Brock got a million-dollar advance to write a hit job on Hillary Clinton similar to the one he'd previously thwacked Anita Hill with. (Brock famously authored the line about Hill: "a bit nutty and a bit slutty.") However, in the course of writing the book he had a conversion experience and renounced the right—he'd previously been a protégé of Tyrrell's at the *American Spectator*—and in 1996 came out with *The Seduction of Hillary Rodham*, an intermittently sympathetic biography. (Most of his enmity is reserved for Bill Clinton, the titular seducer.) Nevertheless, his sustained attention to Hillary's hair, eyebrows, and overall physical appearance—which "had always been something of an Achilles' heel"—reveals a residual tendency to play to his old pals too. (They weren't having any of it; the book tanked.) Thus we hear that in college, Hillary had an average figure and thick legs, wearing the loose-fitting flowing pants favored by the Vietcong. Along with these she sported white socks and clunky sandals (here even I must protest!), wore no makeup, piled her hair on top of her head, and "wouldn't have won any beauty contests." In other words, she came from the "look-like-shit school of feminism," as Brock quotes a source saying. As a young wife in Arkansas (or Dogpatch, as the author likes to call it), she cut a "comic figure," her hair fried into an Orphan Annie perm, with a "huge eyebrow across her forehead that looked like a giant caterpillar." (Note to young women planning future presidential runs: start tweezing now or you'll hear about it later.) Her struggles with her weight were a source of endless amusement to the secretaries in her law firm—Hillary was on a perpetual diet and came to work with bags of lettuce that she nibbled on throughout the day. (Though elsewhere we hear she wouldn't diet.) She wore big khaki skirts and striped blouses because she didn't have the body for a dress; she wouldn't wear panty hose or shave her legs, thus was considered a ball-busting feminist. She also wore unflattering purple glasses. When trying to look good for court appearances, she'd put on awful plastic jewelry and high heels she couldn't walk in.

Finally she succumbed to a makeover during the California primary, hiring the one-named hairdresser Christophe to lighten her hair; a television producer pal got the head of wardrobe from *Designing Women* to devise a more feminine wardrobe to replace her frumpy knit power suits. Still, when left to her own devices she continued to be a fashion disaster, for instance, insisting on wearing a floppy blue velour hat to the inauguration despite the protests of her staff. Brock says that Hillary wasn't a phony and shouldn't have had to play the part to advance Bill's career; he also says that her physical appearance should never have become a political issue (notwithstanding the amount of time he devotes to cataloguing it). But he concludes that her shifting image has undermined public confidence about who she really is. Brock himself is someone who has not always seemed to know who he really is, which could account for the flashes of empathy in his book. After a successful and lucrative stint as resident dirt-digger at the *Spectator*, he ditched his old friends in a flamboyant screed published in *Esquire* titled "Confessions of a Right-Wing Hit Man" and now spends his time going after the right instead, as a media watchdog for a liberal website.

John Podhoretz, in *Can She Be Stopped? Hillary Clinton Will Be the Next President of the United States Unless . . .* (2006), worries that Hillary Clinton is a political juggernaut who can't be defeated, or not unless the right gets its act together. Not as extensive in his attention to anatomical details as some of his cohort (he publicly criticized Klein's *The Truth About Hillary* when it was published), he still manages to observe that Hillary never quite figured out what to do with her hair and clothes, isn't a raving beauty, and has a manner that's almost pathologically unsexy. Interestingly, Podhoretz's worry is that this may actually work in her favor as a presidential candidate, even though it worked against her as First Lady, when she was incapable of acting demure enough for the job description. Yet her very lack of femininity—she's "neither girlish nor womanly" with a "hard to describe style"—may in fact be the perfect blend for the first woman president, since a president has to be a little scary and not seem to be emotional—that is, she should basically be an unlikable bitch. "And Hillary is a bitch," Podhoretz writes. Feigning concern that saying this kind of thing makes him sound sexist, he explains that a woman presidential candidate needs to show that she can be manly, and if ever there was a woman politician who can pass for a tough guy, it's Hillary. In his assessment, this is the only political advantage her feminism offers her. To become president, she has to "play up her antifeminine qualities without being completely without charm and appeal." She needs to have a hard, steely edge, so the cold, unemotional quality she conveys is actually an advantage. (This may not be entirely wrong.)

About that coldness. Among Hillary's critics, her body temperature seems to get almost as much attention as her hairstyle. Brock reports that in high school, her classmates nicknamed her Sister Frigidaire. Tyrrell too thinks she has a cold quality. Klein quotes Richard Nixon, of all people, who says that Hillary is "ice cold." Klein also relates a story about a fight Hillary once had with a college boyfriend over her dislike of skiing, a fight that ended with Hillary retreating into "icy silence." Sounding like a Monty Python rendition of a Freudian analyst (Hamlet's psychiatrist: "Ah! It's the sex, is it?"), Klein speculates that the skiing dispute "might have been a substitute for an honest discussion about Hillary's sexual frigidity." (Get it? *Icy*.) In other words, the dilemma of Hillary's sexuality does seem to hover somewhere in the vicinity of these hair and temperature issues. Klein spends many pages retailing innuendos about her lesbian proclivities (though he also says she'd had a torrid affair with Vince Foster, in addition to being sexually frigid and asexual). Tyrrell thinks she should divorce Bill; Brock thinks she fell so hard for Bill, the master seducer, that she gave up her political principles for him; Podhoretz thinks she has a kind of addiction to Bill. The conspiratorially minded Carl Limbacher, in *Hillary's Scheme: Inside the Next Clinton's Ruthless Agenda to Take the White House* (2003), believes she's nothing less than an abettor to Bill's philanderings, and tried to suppress evidence of them lest it "raise questions about her own private peccadilloes." According to Limbacher, Hillary Clinton is "a victimizer who actually enabled her husband's predations," since she's "married to a ravenous sexual predator at best and a brutal serial rapist at worse," but sticks with him because he's her springboard to power.

This is the rhetoric of male hysteria: endangered and shrill; aggressive, yet also silly. Hillary's critics claim they don't hate her because she's a powerful woman, they hate her because she's *Hillary*. By attacking her, they're just refusing to kowtow to political correctness. They're not threatened! Or if they're threatened, it's because she's a hard-core leftist who wants to radicalize every social institution she can get her hands on. But this has a hysterical edge too: if there's one thing Hillary Clinton isn't, it's a radical. (Ask any leftist.) So what kind of threat does she pose exactly? It's not a question that those exhibiting the symptoms would be capable of answering directly. Political anxieties and sexual anxieties are so entangled at the moment that prying them apart seems doubtful.

HOW HUNGRY IS HILLARY?

READING THE CULINARY CLUES

BY MIMI SHERATON

The question, of course, is authenticity, as it seems always to be where Hillary Clinton is concerned. She wrestled with the issue as far back as 1967, when, as a student at Wellesley, she wrote to an old high school pal, John Peavoy, "There is a smorgasbord of personalities spread before me." A smorgasbord, of course, is a diverse buffet from which one can pick and choose those foods that strike the fancy and suit the mood. So which personality has Hillary selected? Or, in correct Scandinavian tradition, is she still going back for seconds and thirds?

That food preferences are clues to personality was the firm belief of the French gourmand and essayist, Jean Anthelme Brillat-Savarin who in his 1825 work, *The Physiology of Taste*, wrote, "Tell me what you eat and I will tell you who you are." In the absence of an interview with the lady herself, I have pieced together clues as to Hillary's eating habits from various reports, to try to determine whether she is tough and self-assured enough to function as president or merely a food-fashion victim who opts for the flavor of the moment. Does she indulge wild, instinctive cravings with a hat-over-the-windmill bravado, or is she an abstemiously disciplined eater who can be counted on to make the sensibly healthful choice?

In short, would the real Hillary in a private moment go for an olive burger or a Boca Burger? When, in May 2007, Associated Press reporters asked the presidential hopefuls to name the single item that most recalled their back-home origins, Hillary Clinton chose the olive burger served at the Pickwick, her high school hangout in Park Ridge, Illinois. This Greek coffee shop is still in business, next door to a land-marked art deco movie house, also called Pickwick. To find out just what an olive burger might be, I called the owner, George Paziotopoulos, who bought the restaurant from his cousin about eight years ago.

"It's six ounces of grilled ground beef sirloin on a toasted hamburger bun with a thick topping of chopped, pimento-stuffed green olives," he said, pointing out that he was not the owner during Hillary's time. However, in May 2003, he welcomed her back, with Barbara Walters and a local friend in tow, and while filming an interview, they all ate (report-edly "with great relish") what had been renamed the Hillary burger, seasoned with Dijon mustard, a pretty fancy condiment for a Greek diner.

Unable to get to Park Ridge to take advantage of the $6.75 olive burger special, complete with choice of soup or salad and coleslaw or fruit, I cooked up my own version. It was a powerfully brassy, acidic concoction with the merry Christmas touch of the red-and-green olive topping that, along with the mustard, zapped any flavor of beef. Strong stuff for a strong palate, I thought, with a certain respect. Peavoy told me that he remembered the Pickwick, but he did not recall being there with Hillary and so could not report on how many she ate or whether she "scarfed" them down, as has been stated in one account.

Next I interviewed Walter Scheib, who worked as the White House chef for the Clintons (and, briefly, for the second Bushes). Scheib recently published a cookbook memoir, *White House Chef*, which offers many clues to Hillary's preferences. Had she ever asked for an olive burger or a Hillary burger, I asked?

"No, but I always kept Boca Burgers in the freezer," he said, referring to a brand of soy protein patties. "She liked them for snacking." When I obtained some Boca Burgers and pan-grilled them, as directed, they turned out to be miserably limp, grassy-tasting little disks that might have been produced by Rubbermaid. Asked what he thought of them, Scheib replied, "I and my cooks figured they'd be okay if you added lots of cheese and bacon." Not to mention a half-bottle of ketchup and maybe a soupçon of Dijon.

And so, the question remains: How could the lover of the lusty olive burger ever settle for a Boca Burger? Or had the years wrought changes?

I found few reports of Hillary's gastronomic adventures

at Wellesley or whether she frequented the bygone Bailey's Ice Cream Parlor or ever indulged in two Wellesley dining hall specialties, chocolate fudge cake and peppermint candy ice-cream cake. In one of her soul-searching letters to Peavoy, she referred to a "boy from Dartmouth" with whom she had spent a Saturday evening. The boy turned out to be Robert B. Reich, later a Rhodes scholar with Bill Clinton, and still later the president's labor secretary. In a blog he wrote during the summer of 2007, Reich acknowledged that he and Hillary had gone to see the Antonioni film *Blowup*, during which Hillary wanted popcorn with a lot of butter. "A lot of butter. Significant? You be the judge," Reich wrote.

Hints of other preferences and contradictions abound in Scheib's book. To the First Lady's credit, when she hired Scheib away from the Greenbrier, the spa resort in West Virginia, she insisted on having American cooking rather than traditional French fare, wishing to reflect the ethnic diversity of the country and to showcase American food and wine producers. Call it patriotic or merely politically correct, but that was the culinary persona she chose to project.

At his tryout luncheon for the Clintons, Scheib prepared pecan-crusted lamb with morel sauce, and discovered that lamb is Hillary's favorite meat. This in itself indicates a certain palate sophistication, because lamb has a more complex, gamey flavor than easier-to-like beef, veal, or pork. The sweet potatoes Scheib served with it were spiked with red curry paste. The Clintons loved them, and this prompted Scheib to keep assorted hot sauces on hand at all times. It also offers a personality clue: according to research performed years ago by Dr. Paul Rozin and Deborah Schiller at the University of Pennsylvania, humans are the only animals that are thrill seekers, and people who love hot chilies are considered limited risk takers; they are the kind of people who are willing to gamble or ride roller coasters. (One has to wonder what would constitute a limited risk in current geopolitical terms.)

Although many family meals in the Clinton White House were based on fish and vegetables, with a minimum of starches (it was the Atkins and Dean Ornish era, after all), things were different if Bill or Hillary were eating alone. Hillary went for the exotic flavors of the Middle East—baba ganouj, hummus, and tahini. And if President Clinton was on his own for dinner, he invariably canceled the healthful meal that had been ordered for him and asked Scheib to dig into his secret stash of prime meat and grill a twenty-four-ounce porterhouse steak with béarnaise sauce and fried onion rings, evidence that marital cheating can take many forms.

Being a woman, Hillary is expected to cook, something that is rarely demanded of a male political candidate. Once when she was asked if she was good at it, she answered candidly, "I'm a lousy cook, but I make pretty good soft scrambled eggs." Soft scrambled eggs—another indication of a stylish palate, as are omelets and tossed salads, specialties that she copped to on another occasion.

Of all the eating Hillary Clinton has done, none could be more trying than that required along the campaign trail. Accepting, and then relishing, the specialties of a particular constituency—the more ethnic or regional the better—has become part of the American political ritual. "Love me, love my food" seems to be the challenge, while to refuse or, worse, to start eating something and then not finish it, is seen as a flat-out rejection. In June 2007, Eugene Fraise, the Democratic senator from Iowa held a barbecue honoring Hillary at his farm. "We're not going to vote for them if they don't sit down at our table and have coffee," he said.

And not only coffee. Think hero sandwiches, pizza, calzone, ribs, fried chicken, corn on the cob, nachos, fajita-filled tortillas, pancakes, waffles, kielbasa, pierogi, dim sum, knishes, lox and bagels, and more. One well-retailed bit of apocrypha concerning Nelson Rockefeller's run for governor of New York State suggests the consequences involved in committing a culinary gaffe. Handed a plate of blintzes in a restaurant on the Lower East Side, Rockefeller tasted one and said, "These are delicious. What are they called?" Despite being a local laughingstock for a while, he was elected, but probably with much less of the Jewish swing vote than he might otherwise have garnered.

Candidates on the campaign trail also feel pressure to eat competitively, as Hillary discovered in 2000, when she ran for the Senate against Long Island Republican Rick Lazio. A week after Lazio admitted to being less than enthralled with the sausage, peppers, and onion sandwich that he was served at the New York State Fair in Syracuse, the Clintons arrived, and they both sat down to a very public lunch of that same local specialty. "It's great," Hillary announced through a mouthful of the greasy, dripping creation. She even wore a bib.

Bragging about his wife later at a fundraiser in the area, Bill Clinton said, "After the other candidate refused to eat a sausage sandwich there, this one did not." To redeem Lazio's reputation, his spokesman countered: "Growing up in New York and with a name like Lazio, I'm sure our candidate is a bit more acquainted with sausage, not to mention peppers and onions, than Mrs. Clinton." But not acquainted enough, apparently, as history proved.

As Hillary's campaign for the Democratic presidential nomination progresses, so must her waistline, a situation guaranteed to add stress. Judging by various accounts, much of her public snacking in Iowa consisted of sweet and creamy desserts, perhaps another weakness (Walter Scheib reported that while in the White House Hillary, like Chelsea, loved Dove bars). Following news reports, I traced the Clintons' visit to Whitey's Ice Cream shop in Davenport, Iowa, where, via telephone, Jan, the store manager, said that she had witnessed Hillary order a Drumstick—a chocolate-and-chopped-nut-coated vanilla ice cream on a stick—while Bill had peach yogurt in a waffle cone. Asked if they each finished the whole thing, Jan replied, "They sure did!" That was only the first of three caloric pit stops within thirty hours. The couple went on to a Dairy Queen in Nashua, where Bill sipped a strawberry malt while Hillary chose the raspberry, and then they dropped into another DQ near Grinnell, where Bill had a grilled chicken sandwich and, for good behavior, was rewarded with a taste of Hillary's Snickers Blizzard.

For a final bit of insight into food and its meanings, consider the video that the Clinton campaign put out parodying the last episode of *The Sopranos*, in which Tony, Carmela, and A.J. eat onion rings together in a diner. In the Clinton version, Hillary orders carrot sticks instead of onion rings, and when Bill protests, she tells him, "I'm lookin' out for ya." Ann Althouse, a law professor who writes a popular blog from Madison, Wisconsin, conjectured on the deeper meaning of the carrot sticks: "I doubt if any blogger will disagree with my assertion that, coming from Bill Clinton, the 'O' of an onion ring is a vagina symbol. Hillary says no to that, driving the symbolism home. . . . [And] what does she have for him? *Carrot sticks! . . . Here, Bill, in retaliation for all of your excessive 'O' consumption, you may have a large bowl of phallic symbols.*"

In the end, how can anyone not admire a woman who, like so many of us, is torn between renunciation and appetite, with a weakness for the hot and spicy and the cool and sweet, and who surely represents the people's palate?

Significant? You be the judge.

CONFESSIONS OF A HILLARY HAGIOGRAPHER

SEEING HILLARY PLAIN

BY JUDITH WARNER

A hagiography."
That's what a French reviewer called my book, *Hillary Clinton: The Inside Story*, when it was first published, in translation, in the summer of 1993.

The word, taken from the only review the book received upon publication (it was an "instant" paperback, not the kind of biography to get serious attention in the United States), came as a shock. How, I wondered, could anyone say that I'd made Hillary Clinton out to be some kind of saint? Wasn't my book, as its back cover proclaimed, a "probing, fair-minded biography"? Hadn't I shown Hillary to be a complicated figure, a woman who'd made a series of devil's bargains, in her work, in love, and in her public persona? Hadn't I shown her to be (rather devastatingly, I thought), particularly compromised as a progressive?

Page 115: She had changed her last name from Rodham to Clinton.

Page 209: She believed in the doctrine of personal responsibility.

Page 218: She supported abstinence as a means of com-
bating teen pregnancy. ("It's not birth control but self-con-
trol," she'd said, back in 1986.)

Page 219: She was wary of the label "feminist."

You are perhaps now feeling a little bit confused. You
are thinking maybe that the "inside story" of Hillary's compromised liberalism or feminism isn't all that compelling. Or
noteworthy. Or—to the untrained eye—discernible at all.

This tells me that you were not, as I was, back in the fall
of 1992, a freelance researcher for *Ms.* magazine. You were, perhaps, not obsessed with the finer points of what did or did not
make a true feminist or a true progressive or a feminist-progres-
sive role model. You were, perhaps, not as caught up as I was
then with the "culture wars." Pat Buchanan and Rush Limbaugh
and Allan Bloom and Marilyn Quayle. Catherine MacKinnon
and Naomi Wolf and Anita Hill and Susan Faludi.

I lived and breathed all of that back then. I was 27 years
old and I wanted so badly to have a voice in the debate. So
much of the passion—for and against feminism, working
women, the modern world itself—swirling around in that
time came to coalesce, in the election year, around Hillary
Clinton. I began to entertain hopes of writing a book about
what she represented for the American public. Why did she
provoke such passion—such adoration for some, such hatred
for others? Why had she become such a symbol of moral ruin
for the right and so outsized a repository of hope for the left?

"Nobody will give you money to write that book," my then-boss told me one day in early November. He was a journalist turned investment banker who was writing a biography of Ted Turner with reporting help from my husband Max and me. "But tell a publisher that you can write the first biography of Hillary Clinton and that you'll deliver it in time for Inauguration Day," he said. "You'll have a bestseller."

And so I did. Without Hillary's participation, but with the blessing of, and some help from, her entourage, I put together an annotated source list. I divided it up with Max and a researcher. We interviewed night and day—Hillary's old friends, classmates, former colleagues, associates from Arkansas. I set out to try to read every interview Hillary had ever given from the late 1960s to the end of 1992. I read her speeches and her law articles. I all but memorized her previously published quotes.

And I wrote a book that I thought was honest and true.

Hillary herself told me that she thought it was "accurate" when I finally met her one day in the spring of 1999, when I'd gone to hear her speak. "I could actually recognize myself in it," she said. She was warm and friendly. She told me she often recommended my book to people. In a photo taken of us that day, you can see me virtually swelling with pride. I felt like she'd given me the greatest compliment a biographer could ever receive.

And yet—in truth—had she?

Or did she drive the nail right into the coffin on that tiresome old question of hagiography?

I reread the 1993 version of *Hillary Clinton* this summer. (There is a 1999 edition too—a somewhat more sober, less starry-eyed book, which was written when my French publisher requested a post-Lewinsky update.) There are sentences that made me cringe. Like this, on Hillary's childhood: "Hugh and Dorothy Rodham were parents with dreams, and highest among them was to make their children happy."

On her pregnancy: "Soon Hillary grew beautifully huge."

On her efforts to help Bill Clinton take back the state of Arkansas after his 1980 gubernatorial defeat: "Her speeches were so strong, so passionate, that it was impossible not to be impressed by her."

On her relationship with then 12-year-old Chelsea: "Friends say that together they're just like friends."

And on the Clintons' eventual victory in the ugly 1992 campaign: "Faith won out over fear; vision shone through the political smokescreen."

Who wrote those sentences?

What were you thinking? I wanted to shout at my younger, more credulous self, whom I could see so clearly, hunched over my old Ikea desk in the bedroom of our tiny apartment on East 86th Street in Manhattan. I pictured a half-eaten plate of Chinese takeout on my left and a big glass of Diet Coke on my right. I remembered the smell of stress in the room and the aura of fear.

It was my first book. My big break. I had, I think, exactly seven weeks and six days to report and write it.

Get off my back, I heard my old self saying. *I'm doing the best that I can.*

Did I find it a little bit strange back then that all my sources produced almost identical reports of Hillary's generosity, loyalty, faith, and devotion to public service? Yes, I did—the sort of person they portrayed was utterly foreign to me—but I didn't think the depictions were false.

Did I think there was much more to her relationship with Bill than met the eye? Sure, I did. I had no doubt, then as now, that Bill Clinton was Hillary's one vice. I figured he was the beginning and the end of her dark side. But he was also—then as now—what for me made her human.

Then as now, the thing that made Hillary most sympathetic was the fact that she had lived through ugly times, suffered and compromised and continued through it all to get on with her life. This was the thing that made her real for me in a way that no acts of bold and perfect, clean and correct self-creation could ever have done. After all, what kind of real person who interacts deeply with other real people actually lives her life in that way? Isn't the best one can hope for to muddle through—with dignity?

It would have been nice if, at age 27, I'd been purer in outlook, with greater faith in human possibility and loftier notions about relationships and personal transparency. But I wasn't. If anything, I was more cynical and more hard-shelled than I am now. That—along with what I'd learned of her decidedly middle-of-the-road politics—is undoubtedly why, unlike so many of my peers, I've never felt "betrayed" by Hillary Clinton. For me, she never promised more than what she eventually delivered.

There's no question, looking back, that *Hillary Clinton: The Inside Story* was, in 1992, the best that I could do. Despite the saint-making prose. Despite the excessive and unnecessary quotes from admirers that I strung together like cans tied to the back of a car. It's a pity that I marred my original reporting with the unmistakable writing style of a clip job. It's a shame that I traded in my potential writerly authority against the need to attribute every opinion, every utterance, to a named source. It's clear that I didn't feel entitled, back then, to draw conclusions of my own, to pass judgment, to do anything other than catalogue and collate.

As a matter of style, that's unfortunate. But now I also find it sort of touching, even admirable. At 42, I kind of like the 27-year-old girl (yes, "girl"), who was aware of the limits of her knowledge, who preferred to defer rather than to presume. The girl who was—ridiculous though it sounds—too polite to dig up dirt and too dumb to realize that she couldn't be taken seriously without doing it.

I could never get away with being that dumb, or that principled, now. That's why I'm pretty sure I will never again attempt to write a biography. In the content of *Hillary Clinton: The Inside Story*, you get a very clear portrait of me as a young journalist: eager and earnest, squeamish and hesitant, desperate, above all, to take my place among the young women who were then making their names by attempting to make sense of the cultural scene. But in the style—in the fact that the book, very often, sounds absolutely nothing like me—I also now see something more: a very damning exposition of one of my most basic personality flaws.

"It's like you have Stockholm syndrome," a newspaper editor friend once said, when I came out of a book I was co-writing with Howard Dean fulminating against the lying, corrupted, biased, and cruel national press. "You always over-identify." I vehemently disagreed with her when she said it. But with time, I've come to see that it's true.

After publishing *Hillary Clinton: The Inside Story*, I fell into ghostwriting. I came to it more or less accidentally. Yet I eventually found in it a kind of calling. You see, all the weaknesses that made me not so great as a biographer—the excessive discretion, the temerity of voice—were ideal traits to bring to the job of ventriloquist's dummy. As for the porous boundaries—well, they proved invaluable when the task at hand, as a writer-for-hire, was to enter into and inhabit another person's perspective.

I found that I was good at getting a feel for other people's voices. Once I felt the rhythm of another person's voice, I could pretty much carry their tune. I did this with Howard Dean and former *Vogue* editor Grace Mirabella, and others I haven't the right to name. And I think now that, unknowingly, those many years back, I did it with that mass of adoring people I interviewed about Hillary Clinton. Instead of using their voices to punctuate my own, I buried myself in them.

It was a blurring of boundaries. A form of enmeshment. And it passed, in my mind, for my own personal truth.

Today I am indeed paid to write about what Hillary Clinton (among others) represents for the American public. In my columns for the *New York Times*, I've taken up the issues of her sexuality ("The Cleavage Conundrum"), her "mimicry of authenticity" (sorely lacking), her "cuckolding" by Bill, and, repeatedly, why she elicits such gut-level hatred from the kinds of women who should, on the surface of things, most like and identify with her.

But the topic that fascinates me the most remains her image—not so much as it is manufactured and maintained by her campaign handlers, but as it's creatively brought into being by the angry and hopeful, cynical and desiring, passionate and partisan American public. The subject of Hillary as fantasy projection in the public mind is endlessly rich, much more revealing and fascinating and, in a certain sense real, I feel, than any portrait I could ever have compiled as a biographer.

I say this not out of residual defensiveness about my old hagiographic biography. It's rather that fifteen years of reading, writing, reporting, and thinking about Hillary Clinton have left me convinced that no one outside her innermost circle will ever truly know the very private woman—as opposed to the very public phenomenon—that she is.

And this impossibility of knowledge was, is, and will always be the real inside story of Hillary.

CAN YOU FORGIVE HER?

Why Hillary reminds us
of the neighborhood scold

by Dahlia Lithwick

I have long believed that if the story of Hillary Clinton had been a movie on Lifetime Television for Women, we would all be naming our babies after her. There is nothing, it seems, that we women love more than plucky survivors who suffer immensely and triumph as the credits roll. If the Hillary Clinton story were a movie, starring Cheryl Ladd or some other former Charlie's Angel as the reviled First Lady with the cheating husband, who went on to become the first serious female contender for the White House, we'd every last one of us be going door to door for Hillary.

But there's something about the reality of Hillary Clinton, the accommodations she has made and the roles she has played, that leaves many of us cold. The question I can't help asking is: Do we only warm to successful women when they aren't real? Is there something about those women who do succeed spectacularly that leads us to conclude that they are either not women, or not human?

I've been polling women friends for years now on their feelings about Hillary Clinton, and, overwhelmingly, what I hear is that their hearts are unwilling to follow where their heads have led. We all like the *idea* of a Hillary; a working mom who is searingly intelligent, ambitious, and determined, making it on her own terms in a man's world. But I've yet to encounter a woman who adores candidate Clinton, who wants to staff phone banks and wear T-shirts for her. Tepid endorsements, yes. Moderate heebie-jeebies, yes. But passion, devotion, and strong personal identification? Never.

Some of the alienation from young professional women seems to derive from a sense that Clinton isn't human; that she comes across as oddly hollow. I recognize that same blasted-out quality behind her eyes as the one I've seen in Supreme Court justice Clarence Thomas. You cannot serve as a national ideological lightning rod for decades without, in some way, being scorched to the core. What's unclear to me is whether we believe that any woman who's achieved what Clinton has cannot be human or whether we think that no real woman would ever make the sacrifices and compromises she has in order to succeed. Either way, the consensus is that she is cold and dead inside, a political Tin Man without the axe and jaunty hat. And as a consequence of that coldness, we cannot bring ourselves to like her.

If Carol Gilligan is correct and women are fundamentally "relational," we may need to like our candidates as much as we respect or admire them. A friend put it this way: "On a strictly interpersonal level, she doesn't seem like someone I'd welcome as an intimate friend. She seems too calculating, which would make her difficult for me to trust. Power weirds me out because you have to sacrifice privacy and intimacy for it."

It's hard to pinpoint when Clinton went off the rails for the educated working women who should have been her biggest boosters. On paper she's done everything right: she went to good schools and worked hard; she balanced work and family and succeeded at both. And after fighting her way into the Senate, she did precisely what any scrupulous student of "different voice" feminism would do: she studied and listened carefully, then forged legislative partnerships with former conservative opponents, including Bill Frist (on improving technology and medical records), Rick Santorum (on kids and the Internet), and Tom DeLay (on improving foster care).

In their book *What Women Really Want*, pollsters Celinda Lake and Kellyanne Conway reveal that Hillary's issues are all largely women's issues: women care about health care, education, and, increasingly, national security, each of which Clinton has made a priority—often crossing party lines to do so. But for every step forward there has been some subtle misstep, some tendency to get lodged between the rocks and the hard places. Some feminists have found it hard to forgive her for achieving her successes by marrying well, and others cannot get past her decision to stand by her husband following a spectacularly public betrayal. Some feel she has sold out women with her aggressively pro-war stance. And coloring it all is the criticism that Clinton hasn't *really* worked for it the way the rest of us have. In the 2006 race for New York's

Senate seat, white suburban women proved an Achilles' heel, so much so that her opponent, Rick Lazio, almost bested her with a snarky eleventh-hour attack ad featuring an annoyed woman in a kitchen, kvetching, "We started out at the bottom and worked our tushes off to get somewhere. No, but Hillary, she wants to start at the top, you know, the senator from New York." Nasty as it was, it spoke to a demographic that consistently told focus groups that Clinton reminded them of someone shrill and judgmental: a nun, their mother-in-law, or Gladys Kravitz—the nosy neighbor from *Bewitched*.

I last had an opportunity to watch Clinton in person last summer, in Washington, D.C., when she delivered the welcoming address at a June meeting of the American Constitution Society, the liberal legal establishment's answer to the Federalist Society. The speech was about privacy and there was a telling moment when she briefly departed from the text and wryly observed: "I imagine some of you are thinking, 'Privacy? What on earth do I know about *that*?' There has been so little in my own life." She paused for a laugh. "But I have a firm commitment to protecting it for the rest of you."

I have had no privacy but I will fight to protect yours. Oy. Who else but a mother could say such a thing? Perhaps it's that lingering voice of our mothers that makes some women so viscerally uncomfortable with Clinton. Where Hillary may represent something just this side of sweetly Oedipal to male voters, she can trigger in females a feeling fraught with memories of guilt and service, hectoring and scolding, a Mother-knows-best quality that is experienced less as comforting than as a rebuke. Speaking of privacy: Clinton has conflated secrecy with privacy for so long, it seems that she can no longer tell the difference. The contrast between Clinton's clenched insistence that nothing in her personal life warrants scrutiny and Elizabeth Edwards' expansive comments on everything from her fight with cancer to her policy differences with her husband, is astonishing. Edwards may well be Everymother, but in the warmest, most nurturing way. Clinton comes across as the stern mom down the street that all the neighborhood kids fear.

But it's not merely that Clinton can feel like a scold. The more pernicious problem is that we may not believe a woman has really earned her successes unless we have witnessed her suffering first. And one of the paradoxical things about Hillary Clinton is that as much as she has suffered—both at the hands of her bloodthirsty critics and through the bad behavior of her husband—she has largely done so quietly and in private. As women who may be struggling to balance jobs and children, to break through glass ceilings, to gain equal pay and equal respect, we see so little of ourselves in her. There is no messiness in Clinton—no "before" picture. Everything appears to be contained and controlled. And for those of us (all of us?) skidding through our days with the babysitter on speed dial and a Cheerio in our hair, terrified of being found out for the professional frauds that we are, we see nothing of that struggle in Clinton. There is never a moment of self-doubt, never a moment of faking her way through.

It is possible that men see successful men and think "I

want to emulate that," whereas women see successful women and think, "I want to see the safety pins and the hairspray holding all that together." That may be why the pollsters have begun to call us "Oprah voters." We like to see all the crying and the dieting because we are still crying and dieting ourselves. We may be judging Hillary harshly because that's the standard to which we are holding ourselves.

There is another new term emerging in the pollsters' lexicon: "the mother agenda." It's only slightly more respectful than "soccer moms," and it recognizes the undeniable muscle women now exercise in the voting booth. But it also assumes that mothers have a monolithic agenda, and that couldn't be further from the truth. Witness the mommy wars that grow uglier with every passing year. We greet each new article or book on child care, breastfeeding, staying at home, and dropping out with outrage, fury, and judgment. It leads me to wonder whether Hillary—in addition to reminding women a bit of their own mothers—makes us all uneasy about our own mothering.

Consider that Clinton first launched the conservative wing nuts into orbit with her now infamous 1992 remark: "I've done the best I can to lead my life. I suppose I could have stayed home and baked cookies and had teas." By pulling rank as an accomplished lawyer and working mom, she artlessly presaged some of the nastiest rhetoric in the mommy wars, be it Caitlin Flanagan backhanding women who elect to spend a nanosecond away from their children or Linda Hirshman ordering us to get back to our cubicles and limit ourselves to a single child (hopefully of the self-nurturing sort). Decades after women earned the right to vote, we are still too busy judging one another, and ourselves, to judge women candidates without tearing them apart.

The smartest women in politics have learned to express their political goals in terms of an invitation answered rather than a lifelong dream pursued. Laura Bush still gives speeches under protest. Nancy Pelosi came to office only to fulfill a deathbed promise to a friend. Men are called to higher office. Women are meant to tumble backwards into it. Oops! I'm a senator! As a woman you serve and you serve and you serve—and if you are lucky you get to someday serve the president. The women who make their own luck and drive hard for their own success are still the stuff of chick lit or prime-time.

Whether or not we find ourselves able to warm to Hillary Clinton the candidate, it's a useful exercise to contemplate what it is about a smart, diligent, loyal, and effective candidate that somehow makes us anxious. Because if what we really want is a female president who is both warmly emotional and coolly professional; who stayed at home and baked cookies while she made partner at the law firm; who can be commander in chief of the NSA in six-inch Jimmy Choos, we are not likely to have a woman president in our lifetimes, except maybe on Lifetime. (Or on ABC, which put Geena Davis in the Oval Office in the series *Commander in Chief*.) As long as we use other women and their choices as feminist Rorschach tests—templates for having it all and failing—we will see their choices as rebukes of our own. Until we learn to judge female candidates less brutally, we may have to content ourselves with a woman president on television. And it's worth remembering that even she got canceled in a matter of weeks.

MACRAMÉ, ANYONE?

WHAT HILLARY DOES TO UNWIND

BY LAUREN COLLINS

Powerful public figures have always had their pastimes: Alexander the Great raised falcons; Louis XIV collected coins; Kim Jong Il likes horseback riding, cognac, and James Bond movies. Jesus dabbled in deep-sea fishing, and Hermes was a jogger. But American politicians, particularly those with a presidential cut of the jib, are perhaps most given to vigorous occupation, and exhibition, of their free time. At least six chief executives, for instance, have been avid swimmers, including John Quincy Adams, who began most of his days in office with a nude dip in the Potomac. James A. Garfield was a billiards aficionado. Andrew Jackson raised pet mice. In the midst of the last election, an orangey John Kerry windsurfed off Nantucket, while George W. Bush—duded out in a cowboy hat, aviators, and rawhide gloves—cleared brush at his ranch. Both men, incidentally, enjoy mountain biking, but Bush *habla español* badly while Kerry admitted, with disastrous results, to speaking perfect French.

In this respect, Hillary Clinton is a bizzarity—a high-achieving statesperson with few discernible interests other than achieving highly. (Hobbies are endemic to the competitive classes. At a college interview, I remember a fellow applicant asserting that he enjoyed needlework and had, in fact, designed and sewn the very outfit he was wearing.) Hillary claims no extracurricular passions (real or contrived), possesses no attributes (save shantung pantsuits). Little accrues to her except the impression of being an abstemious grind. One of the more interesting revelations of the series of letters to a college pal that recently emerged was her penchant for extra-buttered popcorn. It was astonishing, for many people, to think of her indulging in even as harmless a human urge as the craving for snack food. Her dearth of diversions isn't a woman thing—even prim Laura Bush professes a fondness for margaritas, and Condoleezza Rice, a concert pianist and football nut, once ice-skated competitively. It's possible that Hillary shuns leisure in solidarity with the working classes, but, given her innate solemnity and the unrelentingness of her ambition, being easygoing has long been, for her, an unusually tough proposition. Abraham Lincoln, for instance, played town ball, a forerunner of baseball, and George Washington blew off steam with animal husbandry, but Hillary spends her time attending meetings at the town hall, if not husbanding her husband. Maybe she's a gas after a few glasses of wine— you can imagine her, in her unfailingly awkward girlfriend mode, tipsily confiding a Calvin Coolidge–inspired wish to ride a mechanical bull—but she usually sticks to ice water. When pressed about her interests, Hillary has said that she enjoys cleaning closets, but how much fun can that be, given the eventual whereabouts of Monica's blue dress and the missing Rose Law Firm billing records?

Technically, it's not fair to say that Hillary doesn't have any hobbies; more correctly, she has hardly any, and, for the purpose of making herself accessible to the electorate, they're pretty abysmal. Michael Tomasky writes, in *Hillary's Turn*, his account of her 2000 Senate campaign, that he once ambushed Hillary with a question about what passions, if any, she nurtured outside of policy and politics. To his surprise, she, "with more spontaneity than she'd ever shown on the stump, went off on a random and cheerful string of associations": archaeology and ancient civilizations, modern art, the Peloponnesian Wars, Pericles's funeral oration, Dostoevsky, Hardy, a visit to Australia where an aborigine had showed her a home remedy for gangrene involving boiling tree bark. Assuming that Hillary really does hold these things dear, she comes off as more thoughtful, sincere, and well-rounded than the opportunistic political animal we imagine her to be, cynically shoveling down state fair funnel cakes and pretending to root for both the Mets and the Yankees. (That particular gambit *still* annoys me.) It is extraordinary that she has rarely mentioned these interests before or since. There is the danger, of course, of seeming too highbrow, an impression that, in the wake of Kerry's example, she is smart to avoid. But Hillary's reticence also says something about her brand of feminism: it may be that, even given the seriousness of her interests, she remains afraid of not being taken seriously. Her self-protective instinct—born of her demure, mind-your-own-business Methodism, and inflamed by the exposures of the Starr years—is that, however insignificant a bit of information may be, it is no right of ours to know. Hillary, to her detriment, demands her privacy, even when it costs her nothing to relax it. The good news is that we will probably remain blissfully oblivious as to whether she favors thongs or bikinis.

With the advent of social networking sites as political recruitment tools, no candidate (except John Edwards, who has a gravity problem) can get away without at least cursorily listing some "interests" on his or her profile, but Hillary's are an exceptionally frumpy, unrevealing bunch. While hard-ass John McCain invokes every sport known to man, and family guy Mitt Romney mentions "horseback riding with my wife," Hillary settles for the bland and joyless: "a good book," "history," and "speedwalking." (Could she keep up with Harry Truman, whose preferred pace was 120 steps per minute?) Romney also likes waterskiing, which wouldn't work for Hillary, whose hair causes her enough problems even when it's dry. Her "hidden talent," she writes, ignoring the current rage for Sudoku, is doing crossword puzzles. Her worst habit is chocolate. (*Jeezus*, Hillary.) Her last music purchase was Carly Simon's *Into White*, which might be inspired less by musical affinity than by their shared habit of vacationing on Martha's Vineyard. For "Favorite food to cook" she responds: "I'm a lousy cook, but I make pretty good soft scrambled eggs," a safety line she's been using ever since the cookies debacle. Even the word "lousy" reeks of 1950s uptightness, especially in contrast to Barack Obama's chilled-out fondness for "loaf-ing with the kids." Hillary's unapologetically humorless re-sponse to "sleeping-in time"—"I feel lucky when I can sleep until 7:00 a.m."—makes you wonder if her account has been hacked by Tracy Flick, the heroine of *Election*.

It's surprising that Hillary hasn't ginned up some Every-woman hobbies to make us like her—*like* her, not respect, or admire, or tolerate her—more than we do. I personally qualify as one of those woman voters who are the bane of political scientists: the irrational, emotional citizen whose preoccupa-tion with personality and authenticity causes her to lean to-ward candidates whose views are often counter to her own self-interest. But there is something to be said for authenticity as it correlates to honesty: I worry that, because Hillary won't come clean on the stupid stuff, she will be even more likely to hedge on the things—a health plan, a tax code, a war—that really matter. (Then again, look where Bush and Cheney's transparency about bass fishing and quail hunting has gotten us.) And her toadying, just-us-sistas performance at the 2007 National Beauty Culturists' League luncheon was uninten-tionally but embarrassingly reminiscent of Woodrow Wilson's purported habit of making dialect jokes.

For people who wonder who Hillary really is, her worst liability is the pollster Mark Penn, who, as both her chief campaign adviser and the CEO of the public relations firm Burston-Marsteller (the firm's clients have included Texaco, Philip Morris, and the Argentine military junta), has made a dark art of situational enthusiasm. (As *The Nation* pointed

out, he trumpets Hillary's support for labor unions, for instance, while running a company that advises, on its website,
"Companies cannot be caught unprepared by organized labor's
coordinated campaigns.") Penn, who coined the term "soccer
moms" and the strategy of pandering to them—a new-school
pothole politics—has a new book out, called *Microtrends*. The
logic is garbled, but his main argument is that "by the time
a trend hits 1 per cent, it is ready to spawn a hit movie, a
best-selling book, or a new political movement." Whether the
new niche groups that Penn identifies—commuter couples,
wordy women, Protestant Hispanics, sun haters, DIY doctors,
shy millionaires, young knitters, to name a few—are the new
swing segments, self-fulfilling prophecies, or fictions remains
to be seen. But Penn, when I encountered him at a recent
lunch, refused to say whether or not Hillary fit in any of the
seventy-five categories that he claims constitute the rest of her
America. "You'll have to ask her," he said.

In lieu of any convincing information about her passions,
I consulted a book called *Get a Hobby: 101 All-Consuming Diversions for Any Lifestyle*, to try to figure out some interests
that could work for Hillary. Bunco? Scrapbooking? Ikebana?
Home distillery? Or perhaps, like Dennis Kucinich, as he
wrote in his MySpace profile, veganism or peace? The book
begins with a quiz, designed to answer the question, "What's
your hobby personality?" Filling in the bubbles as I guessed
Hillary might, I learned that she is extroverted, history-lov-
ing, independent, meticulous, and patient. (Bill, who claims to
descend from a long line of recreational watermelon growers,

would have to vouch for patience.) According to the authors, Hillary's perfect hobbies, therefore, are beading, model ship-building, silk screening, docenting, historical reenactment, and miniature war-gaming. As the book's entry for that last one says, "What's more appealing than controlling your own little civilization?"

HILLARY'S UNDERPANTS

THE SAD TALE OF "CLINTONSHA," OR SHE-CLINTON

BY LARA VAPNYAR

Sometime in the 1980s, Ronald Reagan made his famous visit to the Soviet Union. It just so happened that my grandfather had died a couple of years earlier, and my grandmother, finally out of the clutches of her sixty-year-long marriage, was free to fall in love. Ronald Reagan was tall and handsome, he had a full head of hair, he smiled a lot, and he wore clean, perfect clothes. In short, he was everything my grandfather wasn't. The fact that he appeared on TV a lot was a big plus, too, because my grandmother rarely left the house and it would have been problematic for her to be in love with somebody who didn't appear on TV a lot. Since my grandmother couldn't hear very well, she didn't understand anything that Ronald Reagan said, even in translation, so she had to concentrate on his visual splendor. And what never failed to make her heart skip a beat was his scarf. Whenever Reagan appeared outdoors, there was a scarf over his long coat. An elegant, clean, white scarf. The cleanness of that scarf became the source of endless fascination for my grandmother.

"He wore it yesterday, but look, it's perfectly clean today," she would say with a thoughtful expression. "Nancy must wash it every night."

More than two decades have passed since then. The Soviet Union collapsed, my grandmother died, Ronald Reagan finished his second term, then got sick, then died too, but the image of Nancy Reagan washing her husband's scarf every night has somehow stayed in my mind. Even now, I can picture Nancy Reagan the way my grandmother pictured her. A frail, exquisitely groomed, doll-like woman leaning over the sink in the bathroom of their hotel room, soaping the scarf with the tiny square of hotel soap, rubbing one end against the other with her slender fingers, getting her skin chafed in the process. Then hanging the scarf over the shower-curtain rod to dry. I would mentally add a few more items to that line of laundry dangling in the Reagans' hotel bathroom. Reagan's socks (black, polyester) and Reagan's underpants (roomy white briefs).

Throughout the years I told many friends about that mental picture, and they all laughed at the ridiculous image of the president of the United States and his wife washing their clothes in a hotel sink. Somehow, nobody paid much attention to the other part of the image: that it was the president's wife who was washing his clothes. This didn't seem as shocking. Perhaps there was nothing noteworthy in the picture because it was consistent with the ideal image of a First Lady: a real lady, a quiet woman, willing to stand behind her husband no matter what, and happy to wash his underwear when necessary.

My family came to the United States in the middle of Clinton's first term, and while most of our Russian friends were liberal-minded and seemed to be quite happy with Bill, they referred to Hillary in an openly derogatory way. They called her *Clintonsha*, which could be roughly translated as "She-Clinton," and they complained about her being loud, pushy, obnoxious, greedy, deceitful, ugly, frigid, excessively smart, self-centered, ambitious, hardly human. Russians will often say what Americans secretly think. We weren't raised in an environment of political correctness, so most Russians simply don't buy the concept of necessary insincerity. Russian men will unabashedly say that even if they support full equality between men and women, they still believe that men are smarter, steadier, and more courageous than women and better suited to positions of power. And Russian women might be at the men's throats when they hear these sentiments, but at the same time they'd refrain from pledging allegiance to feminism out of fear that associating with the movement would somehow make them less attractive or would be viewed as a confession that they are not happy with their men. Few Americans would admit to having these kinds of thoughts, unless, like Lawrence H. Summers, the former president of Harvard, they make an unfortunate slip; or, as was the case with the real people in the movie *Borat*, they are somehow duped into being sincere.

Many men were annoyed with Hillary as First Lady

because they felt that they had been deceived into falling under a woman's power. They voted for Bill Clinton, and now there was this woman pushing her opinions on him, guiding his actions, making him commit one mistake after another. Some even said that Hillary was the real president, and that she was the more important half of the couple. There were many Russian jokes at the time portraying Bill as a pathetic loser and Hillary as a woman with a whip. In one of them (a milder one), Bill and Hillary were engaged in a rather indifferent sexual act, and Hillary stopped mid-coitus and asked her husband: "Say, Bill, have you ever thought that you'd be fucking the president's wife?"

Interestingly, I heard exactly the same joke about Mikhail Gorbachev and his wife Raisa, another First Lady who was hated for being too visible. But while jokes about Raisa were focused on her domineering ways and didn't mention her physical qualities, Hillary was usually portrayed as unattractive and frigid. Some Russian men even blamed her for Bill's troubles with Monica Lewinsky. Of course he had to look for fun elsewhere; look what the poor guy had in his bed.

The attitudes of Russian women toward Hillary are more interesting and more problematic. Many women have found it hard to identify with her: too cold, too invincible, too domineering. And not attractive enough! "There is something fake about her expression." "I hate her smile." "How can she dress so badly, she can have all the best designers in the world!" "What's wrong with her bras? Can't she buy a nice Victoria's Secret push-up?"

It's true that it isn't easy to match the visual splendor of Bill Clinton, or put on a smile as genuine as Ronald Reagan's, but, still, what is so irritating about Hillary's appearance? She has always looked and dressed like an average American woman. "She has a smile of a she-wolf," one of my Russian neighbors said. I've never seen a she-wolf smile, but I imagine it would look cold, steely, and possibly cruel. "Just look at her hiding behind those stupid glasses," my friend said, referring to a photograph of a younger Hillary. I personally think she looked her best in those. Poignantly unlovely, uncertain of what the future might bring, reserved yet very much alive, vulnerable to pain yet strong enough to absorb pain.

And, of course, it was the pain brought on by the Lewinsky scandal that gave Hillary her greatest boost in popularity. (Similarly, most Russians softened toward Raisa Gorbachev only after she was diagnosed with cancer.) When the Lewinsky story broke, Hillary seemed to turn human overnight. Finally, she gave women something to identify with. She had exactly the same problems that they did. And her unwavering support for her husband, amazingly, made them feel better about themselves. Her attitude even thawed men's hearts a little. Who wouldn't wish for a wife who behaved like that in a crisis? People even forgave Hillary for not being attractive enough or not being able to choose the right bra. Or, rather, they started to find it endearing.

What is disturbing about all this is that Hillary gained popularity by playing the role of the long-suffering wife. A long-suffering wife would wash her husband's underpants in a hotel bathroom. In 2003 Joe Klein wrote a piece in *Time* magazine called "The Humanity of Hillary": "As First Lady, she was a confusing and an uncomfortable public presence—a feminist who came to prominence as a wife, a professional woman laboring under the burden of a dainty, antiquated official title. She was independent, tough-minded and yet allowed herself to endure one of the most spectacular spousal humiliations in history." Klein suggested that perhaps Hillary was best suited to being a senator and shouldn't consider a run for president at all, because "her victory would be easily attributable to her husband's genius—and she knows that the first woman President shouldn't be elected like that."

I hear this sentiment a lot when I discuss Hillary's chances of becoming president with other Russian women. Many of them, who really want Hillary to win, seem to base their optimism on Bill's presumed influence and assistance. "It's okay," they say. "He will help her to win and then tell her what to do." And this is being said by the same people who complained about Hillary guiding Bill's decisions when she was First Lady.

I can't quite reconcile the fact that Hillary arrived at her present status by riding the wave of her husband's enormous celebrity, and that she became so popular by forgiving his cheating. I would prefer a woman who rose to prominence all on her own. But let's be realistic. We are not there yet, and we are not even firmly on the way to getting there. And if the only way for a woman to arrive at a position of true political power is by being a wife, it is not her fault, it is society's fault.

In the end, I don't care how she gets there; I just want her to get there.

Hillary is smart, she is tough, she is steady, and yet she is capable of being flexible. She does seem to be the best candidate for the job. Yes, it might have been Hillary's willingness to wash Bill's underwear that helped her get where she is, but if she gets elected, there will be no turning back: the perception of a woman's role will inevitably change. I wouldn't say that I really want to envision Bill washing Hillary's underpants or her Victoria's Secret bras in a hotel bathroom; I just wish that the image of a woman washing her husband's underwear didn't continue to seem admirable.

"Hillary, the president? No way. She doesn't stand a chance." So my mother's podiatrist, Doctor Losyev, told her during her annual check-up recently. "Why?" my mother asked him. "For one, because no man would agree to be under a woman. Not here, not in America." Doctor Losyev, a fairly successful Russian émigré and a very imposing man in his forties, was kneeling at my mother's feet at that moment, but apparently the irony of the situation was lost on him.

THE SELF-RELIANCE THING

GETTING OVER THE GIRL WHO ALWAYS DID
HER HOMEWORK

BY MARIE BRENNER

**HILTON HEAD, SOUTH CAROLINA,
JANUARY 1, 1998, 6:30 A.M.**
The First Lady of the United States, Hillary Rodham Clinton, is alone on the beach at Renaissance Weekend. In the chilly dawn, she appears from a distance as a melancholy figure bundled in a pink sweatshirt. Her Secret Service detail is nowhere to be seen. Just Hillary walking against a strong wind, locked in thought.

We search public figures for hidden intentions, for motives we can recognize as our own. All that weekend, tension had swirled around POTUS and FLOTUS. The Clintons, usually social and approachable, seemed hidden behind a shield of distractions and defenses. Something big was happening, something bad. Clinton courtiers whispered about the family's mysterious distress. The weekend gathering had mushroomed, by that year, into a tedious scrum of telecom billionaires, Nobel nominees, media climbers, and rich Rolodex fillers puffed with self-regard. The debate topics were as quaint, in retrospect, as the future of the buggy whip: Would the Internet change the way we read? Would Japan collapse? But the main attraction, as always, was the first couple.

The previous night had been the climax of the weekend: Hillary had been scheduled to introduce the president and launch the New Year's Eve festivities. The Clintons were late. And then, really late. Finally, she appeared. Here, in paraphrase, is what she said:

> *All of us have been in the position where we have had to appear in public to introduce someone . . .*

She paused. Did she say she was angry? She was clearly upset. After a moment, she continued:

> *When that person is the president of the United States, the leader of the free world, imagine the situation you are in . . .*

It was unusual to say the least. The First Lady was cranky and detached, performing like a robot. When the president finally appeared, we all scrutinized the manner in which the couple greeted each other. It seemed measured, chilly. Hillary vanished immediately afterward and did not reappear all evening. That night the president, as usual, delighted his audience with his command of the issues, discussing climate change and new challenges posed by the Internet, among other topics. He stayed for hours at the party afterward, drawing energy from the strangers who mobbed him, and was still in the ballroom at 3:00 a.m. talking geopolitics with a high school student, a sea of empty chairs around them.

A few hours later, I spotted Hillary, alone on the wintry beach. What is she doing here, I wondered? We passed without saying a word. Hillary looked away. Her face was a mask of gloom and disconnection.

Days later, the news of Clinton's affair with Monica Lewinsky surfaced and people were predicting that he would be impeached. In the opening weeks of 1998, a Gap dress and a thong would bring on a personal catastrophe that would threaten Hillary's marriage and career and send the nation and the Democratic Party into a tailspin. The Clintons would face the ongoing inquisition of special prosecutor Kenneth Starr, a zealous twentieth-century antagonist out of Dickens. Hillary's deposition would be routine; Bill's would not.

During the investigation, it was sometimes hard to judge which side was more absurd: the creepy voyeuristic Starr or the Clinton camp, with its manufactured talking points, its oppo attacks orchestrated via the basement press pen, its war room, all designed to torch the twenty-four-year-old intern, Monica Lewinsky. Hillary stuck by Bill through the lies and the confessions, the impeachment and the acquittal. Through it all, she summoned her keenest survival skill, the opaque deflection that is the essence of her character.

I'd caught a glimpse of that on the beach at Hilton Head. The image stayed with me for the next decade, and I think of it now, in connection with Hillary's current storyline. As she heads into the race to become the first woman president of the United States, there are rumblings that the coming campaign could turn into 1998 redux. Is Hillary prepared?

After the Lewinsky mess, she maintained that her husband's vulnerabilities had surprised her. Many believe that this public posture put her leadership capabilities in doubt, and that it has contributed to one of Hillary's enduring problems—the fact that she is ardently disliked by alpha women of a certain age. She is understandably fierce about her desire to shroud this complicated aspect of her marriage—a tough stance to maintain in our tabloid culture. There is no debate about her international expertise and her seriousness on the issues. The persistent question nagging some detractors is this: If she routinely lives with this level of denial in her private affairs, how can she lead the free world?

Trying to understand why Hillary seems to put so many mature, educated women over the edge, I recently raised the issue with a lunch table full of New York professional women. One, a literary agent, vibrated with rage: "Why do I hate her? She is a liar! Her marriage is phony! She has stayed in a marriage with someone who is unfaithful! The woman never says an authentic word! And what about the war in Iraq? She never takes a principled stand on anything! How about the flag-burning law?" Others chimed in about her shameless pandering to the red states on the matter of global trade and the resetting of the Chinese yuan.

Soon, the group was buzzing with knowing innuendos.

One woman said she'd heard that Bill—they call him Bill—is still out there trolling, and that you cannot touch this subject with Hillary's advisers without them going nuclear. One woman said, "This is the question that no one in the Hillary camp seems prepared for. What if? What if?" Another woman added darkly, "Defending Hillary is not a growth industry."

But why does this anger attach itself to Hillary and not to Bill? Do they blame her for him? The fact is, many people view the two of them as a unit: in discussions about the Clintons, you often hear the phrase "Bonnie and Clyde." It's a reference to the crime-spree-like string of allegations that follow them—the Marc Rich pardon, the ties to shadowy businessmen for campaign contributions, Whitewater, filegate, travelgate—all Clinton scandals that were perhaps blown out of proportion by right-wing adversaries. The Clintons keep trying to start the national tape in a new place, keeping the conversation on the topic of the first viable woman to run for president. But inside the green zone of the east and west coasts, it's not easy.

I put the question to David Garth, the man credited with helping to invent political TV advertising. Garth started his career with Eleanor Roosevelt and Adlai Stevenson in the 1950s and went on to help elect everyone from Israel's prime minister, Menachem Begin, to three of New York's mayors: Ed Koch, Rudolph Guiliani, and Michael Bloomberg. Once, when Garth was asked if politicians ever really believe in anything, he answered, "They start off with beliefs and they end up . . . hungry."

He repeated my question: "Why do so many women hate Hillary?" His answer came fast: "It's The Voice."

The Voice, he said, grates across our interior chalkboard, reminds us of the fifth-grade teacher we despised. The Voice says, "Look at me. I am the smartest person in the room." There is no seduction in it, no hint of the velvet intellect of Mario Cuomo ("Let me take you on a journey into my thinking"), no Bill Clinton oozing empathy. The Voice is her shield, irony-free, with no lightness or top-spin. Hillary echoes the tones of the partisan women of her childhood, pols like New York congresswoman Bella Abzug, who could and would knock you over with her purse. Hillary's voice, Garth said, has the earnestness of the Wellesley class of '69 and it falls hard on a YouTube-trained ear.

"In America," Garth told me, "the single most important quality in a political candidate is likability. America does not like an expert." He laughed and added, "That is, until it does." There is little doubt in his mind that Hillary Clinton, who likes to be an expert on everything will pull it off and become the first woman president. "That's how it works," he said. "Everything is true until it isn't."

Garth tells his clients that a political speech is a drink with a blind date, a seduction, with the aim of getting them right into bed. He said that his advice to Hillary would be, "Hey, slow down, put some beats in this." The biggest challenge she faces concerns the media consultants' favorite word: presentation. Her presentation needs to lighten up, unrev, stop quoting *Marbury v. Madison*. It needs to stop jabbing at us with a finger.

Next, I sought out the psychiatrist Justin Frank, the author of the book *Bush on the Couch*. In it, Frank attempted to psychoanalyze the president by scrutinizing his every speech and public statement. Not surprisingly, the book offers up childhood causes—cold mother, absent father, dying sister—to explain the defenses and anxieties that Frank believes have caused Bush to act out, burrow in, and go to war.

What does Frank think is at the root of the Hillary antipathy?

"It's the self-reliance thing," he said. "Self-reliance is a complex trait in a woman, and the way Hillary wears hers like a shield has set off a reaction in women her age: they envy her." According to Frank, Hillary, like many smart women, feels the need to defend her intelligence. She seems compelled to show what she knows, as if to ward off aggression. In Arkansas, she preached, taught Sunday school, litigated cases. She is a performer, a straight-A student, the one who always did her homework. How annoying.

Frank continued: "Clinton is comfortable with being pushy. She is comfortable not being a typical woman. She is comfortable being ambitious." From time to time, her consultants attempt to lighten her up, urging her to emphasize her hair problems, domestic hobbies, and assorted girl stuff to play to her base. But for many people, these weird forays into the world of soccer momism—"I garden and clean my closets in my spare time"—come off as a little wacky, as if Margaret Thatcher suddenly banged on about her recipe for blackberry pie. Laser-sharp intelligence is her authentic public persona, and what is wrong with that? No one is surprised when she is on the Senate floor at 4:00 a.m., long after her colleagues have collapsed.

Critics have ridiculed Clinton for her attraction to a series of weird seers, clairvoyants, and past-life progressionists, including one who was allegedly hired to channel Eleanor Roosevelt. Recently she hired another new adviser, the New Age business guru John Kao, who has been called a "counselor of mindful empathy." Kao, who was paid $70,000 for advising Hillary on the global economy during her Senate race, has taught at Harvard Business School. The blogs have been lambasting him for churning out corporate gobbledygook: in his book *The 20 Statements*, he uses the word "innovation" twenty-five times in the first three pages.

A clue to this desperation—and perhaps to the mystery of what gives Hillary her talent for deflection—might be found in her adolescence. According to Carl Bernstein's biography of Clinton, *A Woman in Charge*, her father Hugh Rodham was a tyrant who presided over his family with a fierce authoritarianism, forbidding new clothes, cars, and other items beloved of 1950s teenagers. Hillary's brothers suffered from what Bernstein describes as their father's pathology, but she appeared to bask in his unconditional love. Early on, Hillary learned to tiptoe around her father and his moods. Letters she wrote when she was in college, however, suggest that she was far less oblivious to her father's tempers than she had previously admitted. She complained that she knew he would never let her work in Africa or, for that matter, spend a weekend in New York; and she confided that she was holing up in her bedroom while he yelled at her brothers downstairs. "I feel like I am losing the top of my head," she wrote.

Interpreting family relationships is a tricky business for biographers, who inevitably bring their own experiences to the task. Bernstein zeroes in on the enduring martyrdom of Hillary's mother, Dorothy, whose own mother abandoned her for years when she was a small child. He suggests that Hillary got her survival skills—her talent for deflection—from Dorothy. Bernstein also notes Hillary's intense adolescent relationship with Don Jones, a charismatic young progressive minister, who took her to volunteer in the slums and to hear Martin Luther King Jr. speak. She suddenly felt buoyed by the call to do good. (Jones later counseled Hillary through the Lewinsky scandal.) When Hillary met Bill, in law school, their attraction was intellectual and emotional, based on a shared conviction to service.

Thirty years later, in 2000, the Clintons were marooned in Chappaqua and Al Gore was not taking the president's calls. Despised by his own party, Clinton was in free fall and Hillary clearly resolved that no matter what, she would save herself. At a low moment in the Clinton administration, Hillary took Chelsea off to India to observe firsthand the grueling realities of life in the third world. It was a page out of the Methodist creed taught to her by Don Jones: "Do all the good you can." Hillary needed to be for something, not against it. "The most difficult decisions I have made in my life were to stay married to Bill and to run for the Senate from New York," she later wrote. Still ahead of her was the challenge of learning all about a new state of 20 million citizens, a broken rust belt economy, and, as Bernstein later noted, a New York City press corps that could "spot a rube a mile away." Soon she would defy the pollsters to be elected in a landslide as senator from New York.

Clinton's pathway to the Senate showcased her ability to tack to the winds. Her "listening tour" was a ploy to try to defang those who saw her as a shrill advocate. She became a new, empathic, centrist Hillary, guided by her husband's political deftness. Frustrated in Chappaqua, Bill Clinton began to think about his future and came to the same conclusion Hillary had: he had to regroup by doing good. Soon, he would be in the midst of a new way back to global power, spearheading $10 billion in charitable projects as part of the Clinton Global Initiative. In 2006, at a CGI meeting, the former president, flanked by a 15-year-old African orphan who now runs an NGO helping other children in crisis, repeated a version of Hillary's mantra: "So much public energy is on the negative, defining people in their worst moments. There is not a soul who can be judged if we only looked at the worst moment in their lives. And when I see you all here, I know there is a reason. It makes you happy. And no matter how old you are, you are too old to waste time doing stuff that makes you miserable when you have options that make you happy."

I thought of this not long ago, watching Hillary speak at a $250-a-seat Women for Hillary event that kicked off her New York City campaign. She glowed from her triumphant performance in a recent debate on CNN. Hillary was on a high, the beach walk of 1998 far behind her. She had marshaled a war chest and a raft of endorsements (everyone from Steven Spielberg to a platoon of American mayors and ministers). Later that summer, she swatted Barack Obama as a political naïf, demonstrating her command of international affairs. Some observers felt that the election was hers to lose. There was one challenge remaining: the self-reliance thing, and the question of whether America is ready for it or not.

27

THE ROAD TO CLEAVAGEGATE

WHAT DO WE WANT FEMALE POWER TO LOOK LIKE?

BY ROBIN GIVHAN

The arc of Senator Hillary Clinton's aesthetic life can be divided into five periods, each of them filled with ambivalence and paranoia. To watch her negotiate headbands and black pantsuits, cleavage and cover girl status, is both fascinating and torturous. There is a part of me that simply wishes she would relax into her appearance. She is an attractive woman, after all. (I say that bravely, knowing full well that it is politically incorrect and that it will surely be assigned some untoward motive. But really, it's just a compliment.) Yet each sartorial reincarnation is as prickly and discomforting as the one that preceded it.

Fashion has never been something that most women easily navigate. And many are resentful that they should even have to. But until women find a uniform as eloquent and versatile as a man's business suit, they are doomed to stumble through leggings and gauchos, dirndls and ponchos, until they find their own unique foothold. Fashion is hard, which is why we need fashion icons, those women whose clothes seem to paint a highly detailed, exquisite portrait of their interior life. They remind us why the struggle to find our own personal aesthetic language can be worth the angst and frustration.

We feel almost as if we understood who Audrey Hepburn was by virtue of her ability to select the perfect little black dress and to personify the gamine. We saw her as elegant and graceful, a woman who always seemed to be in control and utterly at ease. If the reality and our fantasy were at odds, it didn't really matter. Hepburn had sketched out a public persona and we embraced it.

Ask a woman, particularly a baby boomer, to name women whose style she admires and Hepburn's name always comes near the top of the list, along with Jacqueline Kennedy. Kennedy looked so smart in her structured dresses and her sculpted bouffant. Both women knew how to pull themselves together for the camera—for the historical record. Maybe we also admired Hepburn for her work with UNICEF and Kennedy for the gracious and cosmopolitan flair she brought to the White House, but it was their style that endures most clearly in our memory.

Modern women like to pretend that style doesn't matter anymore, that feminism and modernity have shattered the old notion that there was value in a woman being able to look effortlessly chic. But the truth is, style continues to matter—for men as well as women—if only because when we see it executed with skill and precision, we can't help but notice how powerful it is.

I think of someone like the Washington fixer Vernon Jordan, who is known for both the company he keeps and for his highly polished appearance. There's power in his Rolodex—or his BlackBerry—but also in his Turnbull & Asser shirts and his Charvet ties. His attire reads as expensive without being flamboyant. He is a black man who travels in circles in which he is a rarity. He will stand out because of the color of his skin, so if all eyes are going to be on him, he makes sure that he also looks like the most important man in the room. He wears his clothes with nonchalance. His self-assured manner is interpreted as charisma. He exudes personal comfort and control. Clothes don't make him qualified and confident; they simply accentuate those character traits.

People who have found their personal style can send a message that carries far beyond the range of their vocal cords. They are able to tell a room filled with people who they are—or at least who they wish to be—without uttering a word. There is great intent in the way that the conservative author Ann Coulter packages herself in a short, little black dress with a mane of Lady Godiva blond hair. Her style is adamantly feminine, body-conscious, and striving toward glamour. It comes across as a direct rebuke to the strain of liberal feminism that has women looking at the fashion and beauty industries like they are misogynistic assaults on the collective self-esteem and intelligence of womanhood. According to this thinking, truly progressive and liberated women are too busy breaking through glass ceilings, filing lawsuits with the EEOC, and decrying the pathologies of the fashion industry to bother with manicures, hair colorists, calorie counting, and the top ten things they should buy for fall.

We came to care so deeply—too deeply?—about Clinton's style because her public life personifies the baby boomer's striving, having-it-all, fashion-as-tyranny brand of feminism. In her headband phase, she was going to be a new kind of First Lady, the sort who would not merely host teas and read stories to schoolchildren. She would not be symbolic. Her most memorable statements would not be made with her inaugural wardrobe. During Bill Clinton's first presidential campaign, the two traveled around the country promising voters a two-for-one deal. Her headband was girlish, of course, but it also called to mind a particular kind of young woman—one who sees herself as serious, disdainful of frivolous subjects: someone who considers appearance something to be managed but nothing to be parsed. There was studied idealism in that headband. She was going to save the world, and she needed to keep her hair out of her eyes in order to do it.

Soon after Clinton settled into Washington, a singular fashion catastrophe set the tone for her pre–Monica Lewinsky White House years. Clinton was on her way to testify before the grand jury in 1996. She wore a coat designed by Connie Fails, one of her favorite dressmakers from Arkansas. As she entered the courthouse and turned her back toward the assembled media, she revealed an elaborate pattern on the back of the coat. It was all swirling lines and silver beads. Immediately the pundits recognized it as a dragon. They could not resist what you know is coming: They called her the "dragon lady."

Maybe she was. Maybe she wasn't. Only one thing is certain. The pattern was not of a dragon. It was abstract. Sometimes a coat is merely a coat, and at other times it serves as a Rorschach test for the Washington establishment.

The third period of Clinton's life—the post-Lewinsky White House years—is represented by her appearance on the cover of the December 1998 issue of *Vogue* magazine. The photograph appeared just after the Lewinsky scandal had peaked and it was clear that the First Lady had decided not to toss her husband's—the president's—belongings onto the White House lawn as so many irate women wanted her to do. She had decided to stand by her man. She was never more popular with Americans than during this period. In August of that year, a *Washington Post*–ABC News poll showed that 64 percent of Americans had a favorable impression of her. Gussied up for the country's premiere fashion magazine in a claret Oscar de la Renta ballgown, she looked like a queen. And folks didn't begrudge her that. Looking spectacular after one has been cuckolded—and with the help of a fleshy, young intern with hair as thick as a velvet curtain—was clearly the greatest revenge.

When the First Lady left the White House and, after much public hemming and hawing and aw-shucks stalling, decided to run for the Senate, she made what I argue was her wisest wardrobe choice. She slipped into a simple black pantsuit. It was flattering and stylish but it was also as innocuous an ensemble as she could wear. It was the equivalent of a man's uniform of a dark suit, white shirt and red tie. It removed the topic of fashion from the conversation and it transformed candidate Hillary Clinton into a well-coiffed blond head talking about unions and depressed wages in upstate New York. She won. In her victory speech, she acknowledged her debt to the suits. Six years later, she won again.

The fifth phase brings me to her presidential campaign and the matter of her décolletage, to that incident in late July 2007 when she stood on the Senate floor talking about the rising cost of higher education and it was possible to see . . . cleavage. It was not the sort of Hooters display that might leave you bug-eyed. It was far more subtle. It wasn't inappropriate, but it was noticeable. It stood out because of the location and because of its owner. The Senate floor demands extreme decorum. Men are expected to be in jackets and ties. And until 1993 a woman was required to wear a dress or skirt. This is not a forum that has shifted its rules at the same speed as the rest of the culture. The Senate never adopted a business-casual dress code. And of course, the senator had been on record for preferring a buttoned-up style of dress. She had spent countless hours campaigning in New York making a concerted effort to steer attention away from anything below the neck.

Was it an intentional display of cleavage or just a neckline that unexpectedly slipped a bit low? I believe it was something in between, that it was a conscious decision to dress in a way that was more open, more feminine, and more appealing to a wider range of voters. Instead of dressing like a no-nonsense, black-is-the-new-black, East Coast power woman, she chose attire more reminiscent of a midwestern principal. A hint of cleavage was a by-product of softening up. With that display, Clinton allowed the eye to drift away from her mouth and the words coming out of it to take in the entire person: sexual, feminine, womanly, confident, strong.

When Clinton's cleavage made its appearance on the Senate floor, I noticed, and I wrote about it in the *Washington Post*. Her campaign protested: acknowledging cleavage draws attention to her gender. Foul! The campaign fired off a letter to feminists and friends of feminists asking for help to stop the assault on women's dignity. How could people help? Send the campaign your dollars, your checks, your IOUs.

We have come to a point in our society where we are seriously pondering the question of whether a woman has the toughness, the guts, the intellect, and the charisma to be the most powerful person in the free world. There are some who believe that if a woman wins the presidency, she will do it fighting off all references to her appearance. She will make her way into the White House in a gender-blind manner. But that seems almost as ridiculous as the politically correct contortions people go through to describe a stranger in a crowded room even as they refuse to mention the most obvious and distinguishing characteristic: race.

No one thinks it diminishes a male candidate when he fluffs up his masculinity by discussing his favorite sports team in an interview. Clinton—or any woman—should not feel aggrieved if she is asked about her favorite designer. To be honest, I'd be much more interested in hearing about the male candidates' shopping habits—Wal-Mart or Target, off the rack or bespoke—than in reading about how their style of basketball playing or their temperament on the golf course offers some insight into their integrity or managerial skills.

The ferocity of protest from the Clinton bunker on the subject of cleavage seemed to be a reflection of women's continued insecurity with how they are perceived on the world stage. The fear seems to be that every comment about a woman's dress knocks a few points off her IQ, lessens her authority, and cheapens her. Is a fashion observation all it takes to throw a smart woman off her game?

For many women, Hillary Clinton is heroic for having cleared a new professional path: She was the first presidential spouse to enter the White House with headbands *and* her own daunting legislative agenda. Now's she's the first serious presidential contender with cleavage. But along the way, it doesn't seem as though she has become any more comfortable with her public face. She still hasn't become what we crave, even need: an icon of female power. We want her to tell us, define for us, what female power looks like at the highest level.

Must Clinton adhere to asceticism in her dress, be so blandly attired that all her clothes say to the public is, essentially, "I am not naked and that's all that matters"? Is it fair for voters to demand that she exude as much machismo as the men who have traditionally occupied the Oval Office while looking as genteel, tasteful, and ladylike as the women who historically occupied the East Wing? Where's the middle ground?

I thought Clinton had power dressing down when she ran for the Senate and wore all those black pantsuits. They allowed her to blend in, which is just what you want to do when you're trying to prove to voters that you can work within an institution and be one of a hundred. But what happens when you're striving to be number one? To stand alone?

For all the energy we spend reassuring ourselves that appearance doesn't matter, we can't resist talking about whether a candidate looks presidential. Every man who has ever been a contender knows what that means: a good suit and a conservative tie. It also helps if you happen to be tall and relatively attractive and have a decent head of hair.

We don't know what a female commander in chief looks like. We have no guide. Clinton doesn't have a template to turn to. But she does have very good hair.

28

COLD SNAP: THE SORCERESS PROBLEM

On women being their own enemies

by Susanna Moore

On a recent trip across the country, I made it a point to ask the women that I met their opinion of Hillary Clinton. Often there was hesitation in their reply, and even chagrin, as if their ambivalence made it difficult for them to compose an answer. In the end, many of them said that they found her distant, or ambitious, or lacking in emotion. Many said that they could not bear the sound of Clinton's voice. (This is, oddly enough, a complaint frequently made by male corrections officers in regard to female prisoners.)

It is a commonplace that a woman who steps beyond the conventional boundaries of her sex is said to be ambitious. Or calculating. Or insincere. Or a lesbian. It is an additional commonplace to point out that these criticisms are not customarily employed against men—it has not been said that Bill Clinton was too ambitious (certainly too reckless, particularly sexually) or that Franklin D. Roosevelt was a homosexual (despite the cigarette holder and the cape), no matter the depth of antipathy felt toward them. In the 1991 Senate confirmation hearings of Supreme Court justice Clarence Thomas, Senator Alan Simpson of Wyoming suggested that law professor Anita Hill had deviant sexual "proclivities," and Senator Orrin Hatch said he believed that Hill's testimony under oath had been inspired by *The Exorcist*, a film about a young girl's demonic possession. Male antagonism toward Hill was not surprising. What was more interesting to me was the rage incited by Hill in women. This is not to say that I have not experienced the disturbing hostility that women sometimes display toward one another. But the hysteria provoked in women by Anita Hill seemed so clearly a displacement—a projection, in psychoanalytical terms—that I was startled. Of course, not all women felt this way, but when I asked those of my women friends who had taken vehemently against Hill why they were so certain of her duplicity, their answers were irrational: she should have known better about men; she *did* know better; she'd been around. That she was a black woman only appeared to exaggerate her assumed treachery: she knew better than most women about these things (pubic hair on Coke cans). Those women who could explain their reasons found her cold, calculating, opinionated, ambitious. In a letter to the Senate committee, Mary Constance Matthies, a Tulsa lawyer, wrote that two or three young women, recent graduates of the University of Oklahoma Law School, had told her that "Ms. Hill was very outspoken with respect to her views. This trait was reported present in Ms. Hill to such excess that these women lawyers characterized her as a 'bitch.'"

I bring up Anita Hill and her detractors because these same complaints are made about Hillary Clinton. Although there has long been unsubstantiated gossip that Mrs. Clinton also has "proclivities," it is readily accepted as fact (and despite the reports of those who actually know her) that she is detached, cold, ambitious, and calculating (why else would she have forgiven that husband of hers?). One of the drawbacks of Hillary's coldness, naturally enough, is that she is irritatingly unknowable. This is very upsetting to people. Elizabeth Kolbert, in a recent piece in *The New Yorker*, bemoaned the fact that despite years of study, Hillary still remains enigmatic to her: "Of course, I was trying to get at the 'real' Hillary. (In the interest of full disclosure, I never even came close.)" Is Clinton any less knowable, I wonder, than her husband? than Donald Rumsfeld? Condoleezza Rice? Pat Nixon? (Perhaps that is a definition of charm: an ability to appear known.)

We know that Jung's archetypal Feminine—along with the feelings of anguish, loathing, and fear of danger that it is said to provoke—is not derived from any actual attributes of women. So how is the antipathy that women have for other women to be interpreted? In simplistic terms, if we accept that consciousness is historically experienced as masculine, then the unconscious (that dark, damp murk) is symbolized as feminine. The Jungian psychiatrist Erich Neumann has written: "The whole life of mankind and assuredly of primitive mankind—and in what high degree all mankind is primitive!—is involved in the struggle against the suction of the unconscious and its regressive lure; and this is the terrible aspect of the

Feminine." It is not the collective ritual struggle to overcome the threatening female archetype that is disturbing—in truth, it is quite understandable—but the profound ambivalence and conflict felt by women toward other women.

In the past, women were seen as so threatening that they were literally made to disappear—thousands of witches were killed throughout Europe and in New England until the late eighteenth century. In the West, women are no longer burned as witches (although we do execute them). On the contrary, some of us elect them to positions of power, and some of us allow them to treat us as doctors and defend us as lawyers, and we have always permitted them to care for our children. We watch Judge Judy on television and, of course, Oprah. Nowadays, we profess to understand that the sorceress exists solely within ourselves—we always knew this instinctively, but it still frightens us.

The feminist theorist Hélène Cixous writes that men have always convinced women to do their dirty work for them: "The greatest crime against women, insidious and violent, is that they have been led to hate other women, to be their own enemies . . . to mobilize their immense power against themselves." If we no longer send women to the stake, having more or less successfully sublimated those particular anxieties, we are still deeply ambivalent. Maureen Dowd, in her *New York Times* column, has described Clinton as "Lady Macbeth in a black preppy headband," and has scolded her with equal disdain for assuming too passive a role, particularly in regard to her seductive husband. Speaking of Hillary's appearance at

Bill Clinton's side during his emotional eulogy at the funeral of Coretta Scott King, Dowd criticized Hillary for her quiescence. She "shouldn't have been up there," Dowd wrote. "That bobbleheaded thing was annoying."

Some of my women friends find the idea of a female gynecologist disturbing (repression again). There are women who caution me to avoid female customs officers, as they tend to be tougher on women travelers. I happen to have a woman gynecologist—my phobias lie elsewhere—but I must admit that I do avoid women customs officers. In women's prisons, the female guards are said to be meaner than their male counterparts, the theory being that as women they must go a long way to prove themselves in a traditionally male occupation. This same argument—that a woman needs to doubly assert herself in a masculine role—has been used to defend Margaret Thatcher and Indira Gandhi, both of whom proved to be as tyrannical and aggressive as any man. I do not accept this argument fully. Given the choice, my heart tells me that it would be better to have a woman as a jailer, or president, than a man, but perhaps I am naïve, putting too much faith in maternal instinct (my particular projection). Certainly I have felt that Hillary Clinton is not particularly maternal nor wifely. For me, there is something lacking in her of the tactile. I don't object in the slightest to her lack of domesticity, but I do long for a certain . . . sensuousness. She seems to exist in a chilly void, despite the presence of both a daughter and a husband. They are hers, but she is not theirs. And not theirs despite her proven loyalty and discretion. It is difficult for me to imagine her in an embrace, motherly or otherwise, although she has evinced passion and even rebellion in the past. Her emotion is not amatory, but political.

Clinton's valedictory address at Wellesley was said to be so infuriating to the school's president that a security guard was dispatched to remove Hillary's clothes and glasses from the bank of the pond where new graduates took their traditional celebratory swim. That Clinton's passion is political seems to get on our nerves. Her refusal to accept a woman's traditional powerlessness, and her inability to resort to the equally traditional feminine means to power—wiles, seduction, unpredictability, sexual appeal, petulance—estranges us even further. She appears (who can know for certain? who *needs* to know?) to have a legion of unappealing traits—her impersonal elusiveness; her seeming lack of temperament; her physical blankness (there is rarely a sense of a body under those bright suits, although I recall a poignant glimpse of thigh in a photograph of a dance on the beach with her smiling husband); an absence of what can only be called a *farouche* and girlish charm. But these stand in contrast to what actually *is* real: her intelligence, strength, discipline, poise. Surely these characteristics are more desirable in a leader than intimacy, allure, coziness.

Proust believed that it was a mistake to think of style as simply an embellishment. To him, style was inseparable from thought and feeling. Perhaps that is what those young women in Idaho and Michigan and Colorado were trying to tell me; they have grown up in a culture in which style (although not the kind of style that Proust had in mind) is everything. In Clinton's charmless insistence on control, she makes demands on us that are unsettling, if not frightening. As Cixous writes: "What does she want? What more can she want? What can she want more of? She wants the most."

29

HELLO, MY NAME IS . . .

WHAT A SURNAME SAYS ABOUT US

BY CRISTINA HENRÍQUEZ

Not long ago, I was attending a friend's reading on a university campus when, during the obligatory mingling, I spotted the dean of the school's creative writing department. I was out of a job at the time and I decided to introduce myself. Bolstering my courage, I approached him. "I'm Cristina Kowalczyk," I said. I stopped. "I mean, Cristina Henríquez. Well, both. I'm Cristina Kowalczyk officially. According to the Social Security office. It's my married name. But I go by Cristina Henríquez for writing purposes. So I guess here I would be Cristina Henríquez." I took a breath as he stared at me, the look on his face somewhere between puzzlement and amusement. "I'm having an identity crisis," I said.

I am half American and half Panamanian. My full maiden name is Cristina Elcira Henríquez. Before I got married, I deliberated at length about whether to change it. Generally speaking, I'm hung up on names. I dedicate a few lines of nearly every short story I write to some aspect of a character's name. For me, names are more than simple monikers; they signify something essential and inbred about the people to whom they're attached, they contain the story of who people are, they say as much about personality as any list of adjectives could. When someone misspells my name as "Christina," for example, it makes me feel as though they're mistaking at a fundamental level who I am. A Christina with an *h* and a Cristina without one are two very different people. So when it came to the decision whether to change my last name, I didn't take it lightly.

I spent months calling married friends, both those who had kept their maiden names after marriage and those who had given them up. I suggested to my fiancé that perhaps he should adopt my last name instead of the other way around. Briefly, I considered hyphenation, but there were too many consonants. And then, right around the time I was going to be married, a number of my parents' friends and friends' parents filed for divorce. I did a quick tally: in almost every instance, the woman had kept her maiden name during marriage. So maybe it came down to superstition, or perhaps it was simply a case of odd timing, but something about seeing all of those marriages fall apart and not wanting mine to end the same way convinced me at last to change my name.

My Panamanian grandfather, who was a writer himself, once said to me, "Always use your last name. Do not be ashamed of it." We were standing in the living room in his house in Panama, the long embroidered linen drapes blowing in the humid afternoon breeze. I have no idea now why it came up or what I said in response. But I remember that moment perfectly. It was the reason that even though I decided to change my name legally upon getting married, I still wanted to use my maiden name, to wield it with pride in the world at large. My solution was to use my maiden name professionally. I would be Cristina Henríquez for all matters having to do with writing. I would be Cristina Kowalczyk in my private life. And I would not understand until later, until times such as when I would try to introduce myself to the dean of a creative writing department, how confusing that would be.

I'm not naïve enough to believe that my confusion is simply about a name. At the heart of the matter is a deeper identity struggle, one that has me trapped between two cultures—American and Panamanian—not quite sure where I fit into either. I was born and raised in the United States. I grew up eating macaroni and cheese, hamburgers, and Little Debbie snack cakes. My parents played their salsa albums on the weekends, but I remained squirreled away in my room listening to Nirvana, Superchunk, and Buffalo Tom. I put aside my homework at night to watch *Beverly Hills 90210*, *My So-Called Life*, and *The Wonder Years*. I thought of myself as American. But when my family and I visit Panama, as we did and still do for a few weeks nearly every summer, there is something of me that undeniably belongs there, too. There's a pulse in my veins, a buzz in my bones, that starts as soon as the plane lands. Part of myself that I don't always pay attention to in my everyday life awakens.

In Panama, people want to label me as American. But in

the United States, people seem eager to view me as more Pan-amanian than not. I'm always treading between both sides. In this, I think Hillary Clinton and I are in the same boat. We reside in different cabins, to be sure, but we're both navigating the same ocean, both trying to find our way to the shore that most feels like home.

By now Hillary's story is famous. When she was born, her parents named her Hillary Diane Rodham. When she married, she kept her maiden name and became known as Hillary Rodham. After Bill Clinton lost his bid for reelection in the 1980 gubernatorial race in Arkansas, his strategists believed that one of the reasons may have been that Hillary not sharing her husband's last name put off voters. During the next campaign two years later, she announced that she would henceforth be known as Hillary Rodham Clinton. That time, Bill won.

I can imagine that by the time she had attended Wellesley, become the first student ever to deliver their commencement address, captured national attention because of that address, and went on to Yale Law School, Hillary must have thought she knew who she was. Until I developed the self-consciousness to know otherwise, I thought I knew who I was; it took a while before I realized the full complexion of my personal history. Hillary's definition of herself changed when she met a young, shaggy-haired man named Bill. In the following years, she became a wife, a mother, and the First Lady of Arkansas. In the process, she changed her name to fit her new role and changed her role to fit her new self.

To me, much of Hillary's adult life has seemed to be about balancing competing impulses—the one that puts her behind the scenes and the one that places her squarely at the forefront. Her biography as the First Lady of the United States emphasizes both that she oversaw the redecoration of the Blue Room and that she championed policy issues like universal health care. In her autobiography, *Living History*, she recounts the moment when her status as "wife of" became all too real. She had ordered some stationery: "I had chosen cream paper with my name, *Hillary Rodham Clinton*, printed neatly across the top in navy blue. When I opened the box I saw that the order had been changed so that the name on it was *Hillary Clinton*. Evidently someone on Bill's staff decided that it was more politically expedient to drop 'Rodham,' as if it were no longer part of my identity. I returned the stationery and ordered another batch." She was still, she made sure to insist, the woman she had been before, a woman who wanted to be seen in her own light, far from the long shadow cast by her husband. She was treading between both sides.

Like me, though, I think Hillary must experience her changing identities as being about something more than a name. I used to joke to friends that I had traveled halfway around the world—from Panama to Poland—when I got married. For me, identity has always been about ethnicity. For Hillary, it's about how she has struggled with and resolved, at different times in her life, what it means to be a woman. Even though we're wrangling with different issues, they're fundamentally about the same thing: how we define ourselves, how those definitions brand us to the world, and how who we are stacks up against who we're expected to be.

I have an advantage over Hillary in this struggle. No one batted an eye when I made the decisions I did about my name. Women today straddle the same personal-professional divide all the time. But Hillary can't do anything without being accused of opportunism. She is held to a different standard. If I have the freedom to redefine myself, why shouldn't she?

In 1999, Panama elected its first woman president, Mireya Moscoso. I've thought a lot about Mireya over the years and about the fact that she was a woman in power in a part of the world known for its *machismo*. She was the wife of a former Panamanian president, Arnulfo Arias. Moscoso had no qualms about invoking her late husband to bolster her campaign, although one thing she did not do was use her married name, Mireya Elisa Moscoso Rodríguez de Arias. She left off her husband's name not as a gesture of independence, however, but as a matter of custom, since she was a widow.

Now, early in Hillary's campaign to become president of the United States, her official website and press materials use the name Hillary Clinton. She's redefining herself yet again, treading between both sides. If Hillary wins, she and Moscoso will each have set a new precedent for their countries. If Hillary wins, the American and Panamanian sides of me will be linked by a new thread. And if Hillary wins, she will have to get new stationery.

BEYOND GENDER

The revenge of the postmenopausal
woman

by Leslie Bennetts

As usual, the pollsters and pundits have missed the point. Endlessly posing the same ill-conceived questions in every survey and interview, they remain fixated on the issue of Hillary's sex. But as any pollster worth his retainer knows, you only get the right answer if you frame the question correctly.

And despite the millions of words expended on the subject, it's not really Hillary's gender that's at the core of our deeply ambivalent and conflicted feelings about her. As we continue to dither over whether the United States is ready to elect its first woman president, the real problem is our own schizoid relationship with female gender roles—and the fact that we don't even recognize the true nature of what's bothering us.

In many ways, Hillary's entire adult life can be read as one long struggle against the stereotypical roles that still pinion women to rigid, simplistic definitions, like butterflies impaled under glass. As a brilliant young woman who came

of age when the burgeoning feminist movement was explod-
ing the nation's understanding of the roles women could play,
Hillary seemed bold and fearless, nationally recognized as a
comer even before she arrived at law school. How many student graduation speakers are featured in *Life* magazine?

Fearless, indeed—until she met Bill, that is. Despite the
thrill of finding a soul mate, she must have known instinctively how dangerous the role he offered her would turn out
to be, both for her personal integrity and for her expanding
potential. And so she fought her fate for years, resisting his
proposals of marriage even as he kept importuning her. She
had a bright future of her own, with lots of options to choose
from; leaving the big leagues to become a wife in Arkansas
and take the backseat to an ambitious spouse hardly seemed
like the best possible prospect.

It was only after Hillary failed the Washington bar exam
that she gave in and retreated to Arkansas to support Bill's
unsuccessful race for Congress. However great her joy on
their wedding day, did she also have a sinking feeling about
what surely lay ahead?

At first she was determined to Be Herself, poor dear—
and that meant thumbing her nose at restrictive clichés. Arkansas in the 1970s was hardly the place for a feisty young
feminist who was twice as smart as most of the good ole boys
whose tobacco-stained fingers controlled the levers of power.
But Hillary had to learn the hard way.

First she refused an engagement ring; then she refused
to change her name. She wore a tuxedo to her best friend's

wedding. Making sport of traditional gender roles was not included on the preferred list of local pastimes, however. Disinclined to indulge her playful tweaking of eternal verities it held most sacred, Arkansas swatted her like a bug, squashing the resistance right out of her. When Bill lost his race for reelection as Arkansas's governor, people blamed his refusenik wife's decision to keep her own name. Although there were other factors, that one really hurt.

So much for the feisty feminist, who promptly folded. The price of defiance was too high, and proud Hillary Rodham became Mrs. Clinton. Bill ran again and was duly reelected. He immediately fixed his sights on the White House; his wife retreated to bide her time. Her reincarnation as Hillary Rodham Clinton would come much later, after Bill became president and she thought it was safe to stick her neck out again, at least partway.

Meanwhile, she had to make money. Bill might be the man, but he was also an ill-paid public servant with a pitiful salary. Hillary was forced to become the family's major breadwinner, and the financial pressures of that conventional male burden contributed to the bad decisions that led to Whitewater and the commodities trading scandal.

At times it seemed as if she couldn't win. Bill had been identified as a promising candidate for the presidency early on, but for Hillary the road to reflected glory was always fraught with peril. Gender expectations lurked like minefields, waiting to blow up in her face at the slightest misstep. Hillary ultimately earned respect in Arkansas as both a lawyer and a hands-on mother, but the national spotlight proved far more unforgiving than the small-town scrutiny she had learned to manage in Little Rock. America was loath to relinquish its old-fashioned expectations of a potential First Lady, and whenever Hillary slipped and revealed the least hint of the impatience she really felt, her patronizing attitude toward traditional female roles caused an ungodly uproar. A woman might be able to get away with playing a nontraditional role if she did so quietly, but she certainly couldn't advertise her defiance, let alone condescend to traditional homemakers.

And yet the Clintons made it to the White House anyway. When they first arrived, Hillary thought she could finally do what she wanted. To her, that meant assuming a leadership role in a matter of vital national policy, despite the fact that she was unelected and accountable to no one except her husband. Although Hillary would subsequently insist on interpreting such problems as the product of "a vast right-wing conspiracy," the twin debacles that forced her back into stereotyped gender roles were actually precipitated by her own and her husband's most salient character flaws. Hillary's self-righteous arrogance contributed mightily to the mess she made of health-care reform; that was the end of a First Lady spearheading an important national policy initiative. And so, instead of morphing successfully into the warrior co-president, Hillary was demoted to chastened wife.

To compound her pain, Bill's incorrigible tomcatting publicly humiliated Hillary during the Monica Lewinsky scandal. But the wronged wife was a role the American people could understand—and sympathize with. The protracted agonies of the impeachment crisis notwithstanding, Hillary's approval ratings soared. Having dreamed of winning respect for her trailblazing leadership, Hillary was forced to face the fact that she earned acceptance only as the long-suffering wife who endured the unendurable with dignified stoicism, a sainted ideal enshrined in the national consciousness by the bereaved Jackie Kennedy.

Through it all, however, Hillary seethed with frustration and thwarted ambition, however mightily she strove to contain them. At times she permitted carefully calculated leaks— even encouraged them, in fact. When I interviewed her during the Whitewater scandal in 1994, she was an embattled First Lady who had never run for office in her life. Given access to her friends, I interviewed quite a few. The first time one suggested that Hillary herself might someday make a good president, the idea seemed laughable. When the second and third friends floated the same seemingly innocent, admiring suggestion, it became clear that this was no mere coincidence. In every other detail, the White House had orchestrated the talking points her friends served up with scripted precision, each one offering me the same carefully rehearsed observations about the First Lady. At the time, the presidential trial balloon seemed so preposterous I gave it little thought except as an example of how well-oiled the White House PR machinery was. Hillary appeared to have about as much chance of becoming president of the United States as Barbara Bush.

But Hillary knew herself—not to mention her untapped potential—infinitely better than the rest of us did. Even so, how was she able to see so far into the future? Few people could claim to have accomplished such a feat, especially when the prize was so unlikely as to be an impossible dream. In retrospect, I can't help but marvel at how, during those dark times, the beleaguered First Lady could have held so ferociously to her own vision for an independent future and a heretofore unthinkable achievement she could one day claim as her own—no matter how many years, or even decades, she had to wait.

Which she did, until Bill was done. When his second term ended, Hillary was permitted to step forward at last, like a hungry lioness who prowls the periphery of the kill, waiting submissively until the male lion devours his fill, before she herself is finally permitted to dine on the sloppy seconds.

And yet some leftovers, like revenge, are best served cold, and Hillary's emergence into the starring role has confounded all expectations. Nobody thought she would win the New York Senate race. Who could have predicted that Rudy Giuliani would drop out, let alone that the Republicans would put up as weak a candidate as Rick Lazio to replace him?

But it has never been a good idea to bet against Hillary; the list of those who have done so and lost is legion, and every battlefield she has passed through on her long march is strewn with casualties. Publicly excoriated for her arrogance and rigidity, she transformed herself into a popular senator who astonished everyone with her newfound talent for collegiality, not to mention deft horse-trading. Who could have anticipated that the GOP's favorite villainess would win the respect and cooperation of her male counterparts, even those on the other side of the aisle?

The Senate was one thing. While hardly in the vanguard of the global gender frontiers, the United States has elected women to the Senate before. It's the idea of a female commander in chief that fries our synapses. Sure, India had Indira Gandhi, Britain had Margaret Thatcher, Israel had Golda Meir, Germany has Angela Merkel, and didn't some of those permissive Scandinavian countries have women leaders at one point or another? But the prospect of putting an estrogen-raddled female into our very own Oval Office to order around all the men in the world's most powerful country—why, it's almost more than a body can bear!

To be fair, we are hardly alone in our consternation; nations all over the world seem to be having a collective nervous breakdown over what to do about women and their increasing refusal to play according to the rules. France was thrown into a complete tizzy last year by the simultaneous revolt of the two leading female players in its own presidential race. First Ségolène Royal, the Socialist Party candidate, elbowed aside her lifelong partner, the father of her four children, François Hollande. As head of the Socialist Party, he wanted to run for president himself, but Royal—who was apparently fed up with his infidelity—decided to dispatch him and run instead. (Fuck you, François.)

The woman on the opposing team was no less defiant. Cécilia Sarkozy, the wife of Conservative candidate Nicolas Sarkozy, publicly repudiated the role she was supposed to play with withering disdain. "I don't see myself as First Lady," she said on national television. "That bores me."

Since this was France rather than the United States, the voters decided to overlook her cool contempt for the prize her husband was struggling feverishly to attain. After all, the French electorate had already forgiven Cécilia's earlier transgression when she absconded to New York with a lover, reportedly in response to her husband's infidelity. (Fuck you, Nicolas.) Why is it that so many women have to wait until their husbands inflict cataclysmic betrayals before the wives finally work up the nerve to say, "The hell with it!" and do what they want to do at last?

Cécilia returned and ran away again three times before finally giving up and resigning herself to home and husband. Although an American woman who publicly carried on like that would have about as much chance of becoming First Lady as a Las Vegas hooker who performed an unscheduled striptease in front of Queen Elizabeth on a state visit, the French shrugged and elected Sarkozy president anyway. That seemed to be the last straw, and the Sarkozys soon announced that they were ending their marriage.

Besides, Cécilia was hardly the first Frenchwoman to yawn at the time-honored role of political wife. When Valéry Giscard-d'Estaing became president, his wife was asked what she wanted most to do as First Lady. "To no longer be one," she said acidly. (Fuck you, Valéry.)

While the French bring a more elegant brand of sangfroid to the enterprise, other nations are also wrestling with such issues. Last year the president of Argentina, Nestor Kirchner, announced that instead of seeking a second term, he would step aside and let his wife, Senator Cristina Fernández de Kirchner, run for president instead. One can only imagine the boudoir negotiations that led to this decision.

Such gallantry was not to be found in Australia, which was arguing over whether Therese Rein, the wife of Labor Party leader Kevin Rudd and a mogul in her own right, should have been forced to sell her $180 million company just because her husband was running for election. The wife always pays the price, but it doubtless helps salve the wound if she gets $180 million in exchange for giving up her own power base.

As complicated as it is to navigate the role of the little woman, however, the situation is even more fraught when a female fails to establish her gender bona fides with the requisite husband and children, who supply the Good Housekeeping seal of approval in any culture.

Women running for office have long been attacked for neglecting their children and being bad mothers, although the same standard never seems to be applied to male candidates who come home only to grab some clean clothes. But in a striking illustration of the "You can't win!" bottom line familiar to so many women, Australia's deputy Labor Party leader, Julia Gillard—single, childless, and presumably unencumbered by any needy human she was failing to service while doing her job—was assailed by right-wing critics as selfish and power-mad. Why? Because she doesn't have children.

Although in this case the old fossil trashing Gillard was

a man, when it comes to gender roles, women can be the most lethal enforcers of all, as demonstrated by Laura Bush when asked whether Condoleezza Rice might one day become president. No, said Mrs. Bush, taking aim as deadly as any professional assassin—because Condi is single and doesn't have a family.

In American politics, it has long been axiomatic that women interested in running for political office fare best if they hold off until advancing age has dispensed with most of their family responsibilities. A postmenopausal woman whose children are grown can't be accused of abandoning the little darlings while she is out soliciting votes, and the pesky issue of her sexuality is likely to have receded by that time, relieving the easily distractible electorate of the onerous burden imposed by the strain of concentrating on her ideas rather than her appearance. As long as she's careful to maintain a no-nonsense hairstyle and to wear dignified suits, the average female candidate over 50 can usually get away with diverting attention away from her gender and onto other issues.

She would be well advised, however, to reassure skittish voters by regularly invoking her gender credentials as a proven breeder. A woman can become Speaker of the House, but Nancy Pelosi has to cloak her authority in gender mufti by describing her ability to order congressmen around as using her "mother-of-five voice." A female can't just be strong and forceful and direct in her decision making; she has to revert to being a mom, which we all know is her primary role anyway.

Hillary tries dutifully to play the game, reciting her gender roles on the campaign trail by describing herself as a mom, as a wife, as a working mother, as the daughter of an elderly parent. But her heart isn't in it. It never has been, not even back when she had an adored child still at home. Although Chelsea was the center of her life, Hillary always refused to put her maternal ardor on display, and people were consequently misled, wrongly assuming that she simply didn't care. In fact, she arguably did a better job of parenting than any of her rivals in either political party. But she's still the one we mistrust.

Now Chelsea's an adult living on her own, and Hillary is 60. The kids do grow up eventually, letting Mommy off the hook. But no matter what her age, a woman's sexuality remains a potential bomb that can be detonated at any time, either by her own unwary miscalculation or by a malevolent saboteur. Ségolène Royal may have been a mother of four whose campaign was dominated by the message "that she will be a nurturing, unthreatening mother-protector," as the *The New York Times* put it—but she was also a definite babe, even at 53, and thus "perpetually judged on the way she looks." Her "flouncy" skirts were a constant preoccupation in the media, but it was the arresting photograph of Royal in a blue bikini that really transfixed *tout le monde.* Nothing like a good hard look at those abs and thighs when you're trying to assess a leader's fitness for national office. Ever wonder how Winston Churchill would have fared in the swimsuit competition?

Back in the United States, Hillary had long since squelched her own nervous tendency to experiment with hairstyles and fashion statements, like an anxious drag queen trying endlessly to get the woman thing right. But then came cleavagegate. Having dutifully covered herself up for years in sexless pantsuits—the female politician's equivalent of the burqa—Hillary made one tiny slip and allowed the public to be reminded that she has—gasp!—breasts. Oh, my.

The *Washington Post*, in the person of an overwrought fashion writer, immediately succumbed to the vapors. "To display cleavage in a setting that does not involve cocktails and hors d'oeuvres is a provocation," she wrote. Being subjected to the awful sight of Hillary's cleavage churned up "the same kind of discomfort as a man exposing himself," she went on. "But really, it was more like catching a man with his fly unzipped. Just look away!"

Talk about prurience being in the eye of the beholder.

Gender being the ultimate hot-button issue, however, the game of gotcha continues to be played with a double standard so exaggerated it would seem like a joke if its consequences weren't so serious. In one poll after another, Hillary Clinton is criticized for being "cold" and "calculating." Although she remains married to her one and only husband, whom she has loyally supported for decades despite the most egregious affronts, her marriage is still regarded as fair game by critics who have long attacked her sincerity and devotion, preferring to interpret more than thirty years of unwavering marital commitment as the product of her deranged ambition to gain the White House.

Now, let's see—what kind of adjectives spring to mind when we consider the marital history of the GOP's front-runner, Rudy Giuliani, who married his first cousin, divorced her, got their marriage annulled, fathered two children with his second wife, cheated on her with flagrant disregard for public sensibilities or private pain, and finally announced to the media that he was getting a divorce before he had even bothered to inform his all-too-visibly devastated wife? And who then tried to bring his mistress to the official mayoral residence where his wife and children were still living?

In all those polls that show Giuliani leading the Republican pack, we hear over and over again that people consider him to be strong and forceful. I missed the part where anyone called him cold and calculating.

Even his son's announcement that he doesn't talk to dear old dad anymore failed to make a dent in Giuliani's approval ratings. Hillary raised a daughter who managed to make it through adolescence and young adulthood without publicly repudiating her mother, getting arrested, or even indulging in the frequent public displays of drunkenness favored by George W. Bush's oft-inebriated spawn. And yet Hillary—the devoted mom who raised a responsible citizen who minds her own business and stays out of the tabloid headlines—is the one who's accused of lacking warmth?

Of course, Republican presidential hopefuls have a long history of such hypocrisy. Newt Gingrich asked his first wife for a divorce while she was in the hospital undergoing cancer treatment and then resisted paying alimony and child support until his church had to take up a collection for his abandoned family. He called his second wife on Mother's Day to ask for a divorce after learning that she had been diagnosed with multiple sclerosis. His third wife was a former congressional aide with whom he had a long-running affair, even as he led the impeachment charge against Bill Clinton for the crime of getting a blow job. Cold and calculating doesn't even begin to cover it.

John McCain was married to his first wife when he met the heiress Cindy Lou Hensley; a month after his divorce was final, he married his blond trophy wife. Then there's Fred Thompson, the 64-year-old goat whose current babelicious spouse is young enough to be his daughter.

Not one of these men has remained married to his original wife. And yet none of them has ever been subjected to the kind of sexual humiliation regularly visited upon Hillary Clinton while she was First Lady, when her enemies were so frustrated by the durability of her marital commitment that they resorted to spreading simultaneous rumors that she was a sex-crazed lesbian (I particularly liked the one about how Secret Service agents had to step over the writhing bodies of Hillary and the actress Mary Steenburgen as they rolled around the floor of the White House) and also that Hillary had an affair with Vincent Foster and then snubbed him, leading to his broken-hearted suicide or (depending on which version took your fancy) to his covert assassination by demonic agents working for—you guessed it—Evita Rodham Clinton, as some called her back then.

Demented lesbo? Faithless ho? Heartless bitch? Cold-blooded killer? Why quibble. All of the above, and no doubt more to boot.

Every once in a while, of course, the body politic must reluctantly contemplate actual questions of substance. Here the double standards are even more breathtaking. Hillary has cared about and worked for the same issues throughout her adult life, and yet she is the one accused of being an unprincipled opportunist, even as her Republican opponents shamelessly flip-flop on everything from abortion rights to the troublesome question of gay personhood. But then, they are men, and men are allowed to betray their spouses and children, lie about their intentions, change their positions and pretend they didn't, and pretty much do whatever else they please, all without being held accountable. Women, in sum, are not.

And when their views on a subject do evolve, all hell breaks loose. Hillary got more grief about voting for the war in Iraq and then turning against it than all the Republican candidates combined have gotten about the innumerable changes in their own positions on important issues. But whatever else might be said about Hillary's 2002 Senate vote in favor of authorizing President Bush to take military action in Iraq, it constitutes a prime example of the pressures imposed on female candidates by their gender—and how those pressures can warp their behavior.

Above all, Hillary—as a woman aspiring to be elected president someday—could not afford to be seen as a weak sister on national defense. No matter what she really thought of Bush and his military hubris, Hillary's fear of appearing soft on terrorism or unwilling to deploy military force overrode all other considerations. Given the horrific bungling of the war in Iraq by the Bush administration, Hillary's failure to oppose his plan at the outset ultimately became the single most damaging misstep she has made in her progress toward the White House. She was one of seventy-seven senators who voted to authorize that resolution, but she has paid the highest personal price of any of them.

And yet through it all, Hillary persists, doesn't she? You can't drive a stake through this woman's heart. After all these years, after all the calumnies and humiliations and defeats that would have crushed a lesser mortal regardless of gender, she always gets back up and carries on.

Fourteen years ago, when I interviewed Hillary for a profile in *Vanity Fair*, I quoted *The Economist* as saying that "she is an enigma to many Americans. . . . In her own way, she is as much a chameleon as her husband. Certainly she is clever, focused, and a role model to millions, but for which role? She has as many of them as she has hairstyles: First Lady, political strategist, lawyer, wife, mother, cookie-baker, health-care reformer, *Vogue* picture-poser, intimidator, charmer."

In the years since, Hillary has added a mind-boggling array of new roles to that list: betrayed wife, former White House consort, long-shot Senate candidate, carpetbagger, unexpected victor, well-respected senator, presidential candidate, and finally—against all odds—Democratic front-runner. (Fuck you, Bill. Anything you can do, I can do better.)

"Which role, indeed?" I wrote in the 1994 profile. "Calling someone a chameleon implies that there is something false about each appearance he or she takes on, as if such different identities must be motivated by an impulse toward subterfuge. Few people seem to consider the possibility that all the facets of the prism might produce an equally true picture, and that the real person they are trying so hard to discern cannot be understood except as the sum total of everything they see, and much they are not privy to as well."

Back then, Hillary's friend Susie May said, "What she's really showing people is that all of us have multidimensional lives. She is living one right in front of us."

All these years later, she's still doing it, although the stakes have risen to the stratosphere. Throughout nearly forty years as a public figure, Hillary Clinton has stubbornly refused to let herself, or her life, or her potential be defined by our society's ridiculously circumscribed female gender roles. The prison of those expectations couldn't accommodate the complexities of Hillary even as a dewy-fresh Wellesley graduation speaker, let alone as the battle-scarred veteran of innumerable wars—many public, many more private—that she has long since become.

How did we ever imagine that the one-dimensional roles we force women into could hold her now? For she contains multitudes. The larger truth is, so do we all, if only we are honest enough to acknowledge our true complexities. Isn't it time for all of us to stop expecting people of either gender to conform to simple-minded, anachronistic cartoons? They have never helped us to understand Hillary Clinton, except insofar as we saw her struggle to escape or enlarge them. And they will never encompass everything she is, let alone all she might become, if only we were willing to give her the chance to show us what it really means to be a woman.

LESLIE BENNETTS is the author of *The Feminine Mistake: Are We Giving Up Too Much?* and a longtime contributing editor at *Vanity Fair* magazine.

MARIE BRENNER is the writer-at-large at *Vanity Fair* magazine. She is the author of six books, including *Great Dames: What I Learned from Older Women*, and the forthcoming *Apples and Oranges/My Brother and Me, Lost and Found*.

ROZ CHAST's cartoons regularly appear in *The New Yorker*. Her most recent book is a collection of cartoons entitled *Theories of Everything*. She lives in Connecticut with her family.

LAUREN COLLINS is an editor and writer at *The New Yorker*.

SUSAN CHEEVER is the author of twelve books, including five novels and two biographies; she teaches in the Bennington Writing Seminars and at the New School and lives in New York City.

ROBIN GIVHAN is a staff writer at the *Washington Post*, where she covers the fashion industry and writes a weekly column on culture. In 2006, she won the Pulitzer Prize for Criticism. She lives in New York City.

KATHRYN HARRISON's most recent novel is *Envy*. She lives in New York.

CRISTINA HENRÍQUEZ is the author of *Come Together, Fall Apart*, a collection of eight stories and a novella. Her stories have appeared in *The New Yorker, Ploughshares, Glimmer Train*, and other journals, and she was featured in *Virginia Quarterly Review* as one of "Fiction's New Luminaries." She is currently working on her first novel, *The World in Half.* She lives in Chicago.

LAURA KIPNIS is the author of *The Female Thing: Dirt, Sex, Envy, Vulnerability*, and *Against Love: A Polemic*. She teaches at Northwestern University.

ELIZABETH KOLBERT has been a staff writer for *The New Yorker* since 1999. Prior to that, she was a reporter for the *New York Times*. She lives in Williamstown, Massachusetts, with her husband and three sons.

JANE KRAMER is the European correspondent of *The New Yorker* and the author of nine books, including *Europeans, The Politics of Memory, The Last Cowboy*, and *Lone Patriot*. She has won, among other awards, the National Book Award. Two years ago she was named a Chevalier de la Légion d'Honneur, in France.

SUSAN LEHMAN, a writer, editor, and lawyer, is the co-author, with Edward W. Hayes, of *Mouthpiece: A Life In—and Just Outside—the Law*, with an introduction by Tom Wolfe. Her writing has appeared in publications ranging from the *New York Times* and *The Atlantic Monthly* to *Glamour* magazine.

ARIEL LEVY is a contributing editor at *New York* magazine and the author of *Female Chauvinist Pigs: Women and the Rise of Raunch Culture*. She speaks regularly at colleges and universities about gender politics.

DAHLIA LITHWICK, a senior editor and legal correspondent for *Slate*, writes the column "Supreme Court Dispatches."

PATRICIA MARX is a novelist, screenwriter, and journalist; she writes *The New Yorker*'s "On and Off the Avenue" column. Her novel *Him, Her, Him Again, The End of Him* was published last year. She lives in Manhattan.

REBECCA MEAD has been a staff writer at *The New Yorker* since 1997. She is the author of *One Perfect Day: The Selling of the American Wedding*.

DAPHNE MERKIN is the author of *Enchantment*, a novel, and *Dreaming of Hitler*, a collection of essays. She was a staff writer at *The New Yorker* for five years and is currently a regular contributor to the *New York Times Magazine* and *Elle*.

LORRIE MOORE is the author of six books of fiction, and she teaches at the University of Wisconsin. Her nonfiction pieces have appeared in the *New York Review of Books*, the *New York Times*, the *Yale Review*, and *The New Yorker*.

SUSANNA MOORE's latest novel is *The Big Girls*. She lives in New York City.

SUSAN ORLEAN is a staff writer at *The New Yorker*. She is the author of *The Orchid Thief*, among other books, and she is currently at work on a biography of the dog movie star Rin Tin Tin. She lives in Columbia County, New York.

LETTY COTTIN POGREBIN is a founding editor of *Ms.* magazine, a cofounder of the National Women's Political Caucus, and the author of ten books, most recently the novel *Three Daughters*. She is a past president of the Authors Guild.

KATHA POLLITT is a poet, an essayist, and a columnist for *The Nation*. Her most recent book is a collection of personal essays, *Learning to Drive and Other Life Stories*.

KATIE ROIPHE is the author, most recently, of *Uncommon Arrangements: Seven Portraits of Married Life in London Literary Circles 1910–1939*. She teaches in the Cultural Criticism and Reporting Program of New York University.

MIMI SHERATON, the former food and restaurant critic

for the *New York Times*, is the author of sixteen books, including *Food Markets of the World*, *The Bialy Eaters*, *The German Cookbook*, and, most recently, the memoir *Eating My Words: An Appetite for Life*. She was born in Brooklyn and has lived in Greenwich Village for the past sixty-two years.

LIONEL SHRIVER is the author of eight novels, including *We Need to Talk About Kevin*, which won the British Orange Prize for Fiction in 2005, and, most recently *The Post-Birthday World*. She is the chief fiction reviewer for the *Economist* and the *Daily Telegraph*, and writes features, reviews, and opinion pieces for the *Wall Street Journal*, the *New York Times*, the *Washington Post*, the *Australian*, and the *Guardian*, among many other publications.

DEBORAH TANNEN is professor of linguistics and university professor at Georgetown University. Among her twenty books are *You Just Don't Understand: Women and Men in Conversation*, *Talking from 9 to 5: Women and Men and Work*, and, most recently, *You're Wearing THAT? Understanding Mothers and Daughters in Conversation*. She also writes essays, poems, and stories. Her first play, *An Act of Devotion*, is included in *Best American Short Plays 1993–1994*.

JUDITH THURMAN is a staff writer at *The New Yorker*. She is the author of *Cleopatra's Nose*, a collection of pieces originally published in the magazine; *Isak Dinesen: The Life of a Storyteller*, and *Secrets of the Flesh: A Life of Colette*. She lives in Manhattan.

LARA VAPNYAR, emigrated from Russia to New York in 1994 and began publishing short stories in English in 2002. She is the author of a collection entitled *There Are Jews in My House* and *Memoirs of a Muse: A Novel*. She lives on Staten Island, in New York City.

JUDITH WARNER is a contributing columnist for the *New York Times* and the author of *Perfect Madness: Motherhood in the Age of Anxiety*. Her first book, the bestselling *Hillary Clinton: The Inside Story*, was published in 1993.

AMY WILENTZ is the former Jerusalem correspondent for *The New Yorker*. She's the author of *The Rainy Season: Haiti Since Duvalier* and *Martyrs' Crossing*, a novel. Her most recent book is *I Feel Earthquakes More Often Than They Happen: Coming to California in the Age of Schwarzenegger*. She teaches literary journalism at the University of California, Irvine, and lives in Los Angeles with her husband and three sons.